Lecture Notes of the Unione Matematica Italiana 14

For further volumes:
http://www.springer.com/series/7172

Marco Fontana • Evan Houston • Thomas Lucas

Factoring Ideals in Integral Domains

Springer

UMI

Marco Fontana
Università degli Studi "Roma Tre"
Dipartimento di Matematica
Rome, Italy

Evan Houston
Thomas Lucas

University of North Carolina
Mathematics and Statistics
Charlotte, NC, USA

ISSN 1862-9113
ISBN 978-3-642-31711-8 ISBN 978-3-642-31712-5 (eBook)
DOI 10.1007/978-3-642-31712-5
Springer Heidelberg New York Dordrecht London

Library of Congress Control Number: 2012947649

Mathematics Subject Classification (2010): 13AXX, 13CXX, 13GXX, 14A05, 11AXX

Printed on acid-free paper

Springer is part of Springer Science+Business Media (www.springer.com)

Preface

A classical generalization of the Fundamental Theorem of Arithmetic states that an integral domain is a principal ideal domain if and only if each of its proper ideals can be factored as a finite product of principal prime ideals. If the "principal" restriction is removed, one has a characterization of (nontrivial) Dedekind domains. The purpose of this work is to study other types of ideal factorization. Most that we consider involve writing certain types of ideals in the form $J\Pi$, where J is an ideal of some special type and Π is a (finite) product of prime ideals. For example, we say that a domain has weak factorization if each nondivisorial ideal can be factored as above with J the divisorial closure of the ideal and Π a product of maximal ideals. In a different direction, we say that a domain has pseudo-Dedekind factorization if each nonzero, noninvertible ideal can be factored as above with J invertible and Π a product of pairwise comaximal prime ideals. In each of these cases, if the domain in question is integrally closed, then it must be a Prüfer domain. While this implies, as is often the case in multiplicative ideal theory, that Prüfer domains play an important role in our study, we do provide non-integrally closed examples for each of these types of ideal factorization. On the other hand, we also consider domains in which each proper ideal can be factored as a product of radical ideals, and such domains must be almost Dedekind (hence Prüfer) domains.

This volume provides a wide-ranging survey of results on various important types of ideal factorization actively investigated by several authors in recent years, often with new or simplified proofs; it also includes many new results. It is our hope that the material contained herein will be useful to researchers and graduate students interested in commutative algebra with an emphasis on the multiplicative theory of ideals.

During the preparation of this work, Marco Fontana was partially supported by a Grant MIUR-PRIN (Ministero dell'Istruzione, dell'Università e della Ricerca, Progetti di Ricerca di Interesse Nazionale), and Evan Houston and Thomas G. Lucas were supported by a visiting grant from GNSAGA of INdAM (Istituto Nazionale di Alta Matematica).

Rome, Italy
Charlotte, North Carolina
May 2012

Marco Fontana
Evan Houston
Thomas G. Lucas

Contents

Chapter 1
Introduction

Abstract In this introductory chapter we introduce several variations on factoring ideals into finite products of prime ideals. For example, a domain has radical factorization if each ideal can be factored as a finite product of radical ideals. Such domains are also known as SP-domains. A domain has weak factorization if each nonzero nondivisorial ideal can be factored as the product of its divisorial closure and a finite product of maximal ideals. If one can always have such a factorization where the maximal ideals are distinct, then the domain has strong factorization. Finally, a domain has pseudo-Dedekind factorization if each nonzero noninvertible ideal can be factored as the product of an invertible ideal and a finite product of pairwise comaximal prime ideals with at least one prime in the product. In addition, if each invertible ideal has such a factorization, then the domain has strong pseudo-Dedekind factorization.

The property that every proper ideal of an integral domain can be written uniquely as a product of prime ideals can be taken as a definition of a *Dedekind domain*, which is classically defined via E. Noether's Axioms ([54, Theorems 96 and 97] or [34, Theorems 37.1 and 37.8]). More precisely, a classical theorem by Noether in 1927 [66] shows that an integral domain is Dedekind if and only if every proper ideal is uniquely a product of maximal ideals. In 1940, K. Kubo [55] proved that, if every proper ideal can be factored uniquely as a product of prime (but—a priori—not necessarily maximal) ideals, then the integral domain is still Dedekind. A few years later, K. Matsusita in 1944 [62] showed that, in the previous characterization theorems of Dedekind domains, the assumption of uniqueness can be omitted. This result was implicitly contained in a paper by S. Mori in 1940 [64]. Mori considered factorization properties in rings with zero divisors. He defined a ring to be a *general ZPI-ring* (or, a *general Zerlegung Primideale ring*) if every proper ideal can be expressed as a product of prime ideals. A *ZPI-ring* is a general ZPI-ring in which every proper ideal can be uniquely expressed as a product of prime ideals.

M. Fontana et al., *Factoring Ideals in Integral Domains*, Lecture Notes of the Unione Matematica Italiana 14, DOI 10.1007/978-3-642-31712-5_1,

In 1964, H.S. Butts gave a rather different characterization of Dedekind domains. He declared a proper ideal I of a domain R to be *unfactorable* if whenever $I = AB$ for ideals A and B, then at least one of A and B equals R (and the other equals I). Using this notion, he showed that R is a Dedekind domain if and only if each nonzero factorable ideal can be expressed uniquely as a finite product of unfactorable ideals [12, Theorem].

Essentially, all of the factorizations we will consider are variations and generalizations on factoring each ideal as a finite product of prime ideals. The first variation we will consider for factoring ideals is to simply replace the prime factors with radical ones. We say that an ideal I has a *radical factorization* if there are finitely many radical ideals $J_1, J_2, \dots, J_n$ such that $I = J_1 J_2 \cdots J_n$. In a 1978 paper, N. Vaughan and R. Yeagy [75] studied the rings for which every proper ideal has a radical factorization. They referred to such a ring as an *SP-ring*, for a more descriptive name we will also say that such a ring has *radical factorization*. One of their main results is that a domain with radical factorization is an almost Dedekind domain [75, Theorem 2.4]. An example in [14] shows that there are almost Dedekind domains that do not have radical factorization. The first complete characterization of SP-domains is due to B. Olberding [69, Theorem 2.1].

For a different approach, we can first put a restriction on the type of ideal that we wish to factor, and then ask that these ideals have some "nice" factorization property. For example, a generalized Dedekind domain is a Prüfer domain R such that PR_P is principal for each nonzero prime ideal P and each nonzero prime is the radical of a finitely generated ideal (see Sect. 3.3 for the original definition of generalized Dedekind domains). In terms of factorization properties, a Prüfer domain R is a generalized Dedekind domain if and only if the set of divisorial proper ideals consists of the ideals that can be factored as the product of an invertible ideal (which may be R) and a finite product of comaximal prime ideals (see [31, Theorem 3.3] and Theorem 3.3.6). We will also revisit the notion of factoring families for almost Dedekind domains. Here one restricts to only factoring finitely generated ideals. A factoring family (when it exists) is an indexed set of finitely generated ideals $\{J_\alpha \mid J_\alpha R_{M_\alpha} = M_\alpha R_{M_\alpha}, M_\alpha \in \text{Max}(R)\}$ such that each finitely generated nonzero ideal is a finite product of powers of the ideals from the factoring family, with negative powers allowed. It is unknown whether or not every almost Dedekind domain admits a factoring family.

In all of the other variations we consider, we will allow one factor to be other than a prime ideal, but for this factor we would like it to have some special feature to link it to Dedekind domains. Each nonzero prime ideal of a Dedekind domain is maximal and each nonzero ideal is divisorial, so for an arbitrary domain R we could restrict our factorization restrictions to just those nonzero ideals that are not divisorial. We introduced two such factorizations in [19]. We say that a domain R has *weak factorization* if for each nonzero nondivisorial ideal I, there are finitely many maximal ideals $M_1, M_2, \dots, M_n$ (not necessarily distinct) such that $I = I^v M_1 M_2 \cdots M_n$ (where $I^v = (R : (R : I))$). If such a factorization always exists with the maximal ideals $M_1, M_2, \dots, M_n$ distinct, we say that R has *strong factorization*. This definition varies somewhat from the original definition in [19].

There the M_i were further restricted to being those nondivisorial maximal ideals M_i where $IR_{M_i} \neq (IR_{M_i})^v$. Here we introduce the term *very strong factorization* when the factorization is unique in the sense that the M_i are restricted to being the distinct nondivisorial maximal ideals for which $IR_{M_i} \neq (IR_{M_i})^v$. As we will see, if R is a Prüfer domain, then it has very strong factorization if and only if it has strong factorization. We leave open the question of whether very strong factorization and strong factorization are equivalent for all domains.

Each nonzero ideal of a Dedekind domain is invertible, so another variation would be to have each nonzero ideal factor as the product of an invertible ideal (perhaps the domain itself) and a finite product of pairwise comaximal prime ideals with at least one prime in the product. We will say that a domain has *strong pseudo-Dedekind factorization* when this occurs. For a less restrictive factorization, a domain has *pseudo-Dedekind factorization* if each noninvertible ideal factors as a product of an invertible ideal and a finite product of pairwise comaximal prime ideals. It seems rather remarkable that a domain R has strong pseudo-Dedekind factorization if and only if each nonzero ideal factors as the product of a finitely generated ideal (that may be R) and a finite product of (not necessarily pairwise comaximal) prime ideals with at least one prime in the second factor [70, Theorem 1.1]. However, the equivalence becomes clear once it is known that the latter property implies the domain in question is a Prüfer domain.

In Chap. 2, we collect many of the definitions, properties, and results that we will need in the following chapters, often improving the "classical" statements or simplifying their proofs. In particular, we consider h-local domains, various sharpness and trace properties (definitions are recalled later) and we discuss their interrelations, with particular attention to the Prüfer domain case and to ideal factorization.

There are several types of factorizations that we have not included. For example, in 1964, R. Gilmer showed that a one-dimensional integral domain R is an almost Dedekind domain if and only if each primary ideal is a power of its radical [32, Theorem 1]. Also, D.D. Anderson and L. Mahaney have characterized the domains for which each nonzero ideal can be factored as a finite product of primary ideals [3]. A domain with this property is referred to as a *Q-domain*. In 2002, J. Brewer and W. Heinzer investigated domains for which each nonzero ideal can be factored as a finite product of certain restricted types of pairwise comaximal ideals [11]. They consider three different sets for the comaximal factors: (a) ideals with prime radicals, (b) primary ideals, (c) ideals that are powers of prime ideals. They established characterizations for each of these three sets. In 2004, C. Jayram [53] considered the following generalization of the Anderson–Mahaney factorization into primary ideals: "For what domains R is it the case that R_M is a Q-domain for each maximal ideal M?" An almost Dedekind domain that is not Dedekind has Jayram's local Q-domain property, but is not a Q-domain.

Chapter 2
Sharpness and Trace Properties

Abstract In this chapter we collect many of the definitions, properties and results that we will need in the following chapters, often improving the "classical" statements or simplifying their proofs. In particular, we consider h-local domains, various sharpness and trace properties (definitions are recalled in the present chapter) and we discuss their interrelations with particular attention to the Prüfer domain case and to ideal factorization. Special care has been given to the attributions of the results and to the citations of the original references.

2.1 h-Local Domains

We start by recalling some standard properties of valuation domains.

Lemma 2.1.1. *Let V be a valuation domain with quotient field K, where $V \neq K$, and maximal ideal M.*

(1) Either M is invertible or $M^{-1} = V$ (equivalently, $M = M^2$). Therefore, M is divisorial if and only if M is invertible (or, equivalently, finitely generated or, equivalently, principal).
(2) Let I be a nonzero ideal of V. Then I^{-1} is a subring of K if and only if I is a noninvertible prime ideal of V.
(3) If P is a noninvertible prime ideal of V, then $P^{-1} = V_P = (P : P)$.
(4) If I is a nonzero ideal of V, then $(I : I) = V_P$, where P is the prime ideal of V of all the zero divisors in V/I. In particular, if Q is a nonzero primary ideal of V and $P := \sqrt{Q}$, then $(Q : Q) = V_P$.
(5) If M is finitely generated (or, equivalently by (1), divisorial), then every ideal of V is divisorial. If M is not finitely generated, then $\{xM \mid 0 \neq x \in V\}$ is the set of nondivisorial ideals of V.

The proofs of the statements collected in Lemma 2.1.1 can be found in [24, Corollary 3.1.3, Proposition 3.1.4, Corollary 3.1.5, Lemma 3.1.9, and Proposition 4.2.5].

M. Fontana et al., *Factoring Ideals in Integral Domains*, Lecture Notes of the Unione Matematica Italiana 14, DOI 10.1007/978-3-642-31712-5_2,

Remark 2.1.2. We note that some of the statements of Lemma 2.1.1 have appropriate generalizations to the case of a Prüfer domain.

(1) The first part of statement (1) holds for any Prüfer domain by [24, Corollary 3.1.3] or [51, Corollary 3.4]; i.e., *if M is a maximal ideal of a Prüfer domain R, then either M is invertible or $M^{-1} = R$; in particular, a maximal ideal M of R is invertible (or, equivalently, finitely generated) if and only if it is divisorial.*
(2) For statement (2), the situation is more complicated, and a relevant result will be recalled later (see Theorem 2.3.2(1))
(3) A global version of Lemma 2.1.1(3) will be stated in Theorem 2.3.2 (see both (2) & (3)).
(4) The Prüfer analogue of Lemma 2.1.1(4) is the following [24, Theorem 3.2.6]: *Let I be a nonzero ideal of a Prüfer domain, $\{M_\alpha\}$ the set of maximal ideals of R containing I, and $\{M_\beta\}$ the set of maximal ideals of R that do not contain I. For each α, let $Q_\alpha R_{M_\alpha}$ be the prime ideal of all zero divisors of $R_{M_\alpha}/IR_{M_\alpha}$. Then $(I : I) = (\bigcap_\alpha R_{Q_\alpha}) \bigcap (\bigcap_\beta R_{M_\beta})$.*
(5) Nowadays, much is known about divisoriality in Prüfer domains, but our knowledge remains incomplete.

We recall that a *general ZPI-ring* is a ring in which every proper ideal can be expressed as a product of prime ideals. It is known that such a ring is Noetherian and, at most, one-dimensional [76, Lemma 1 and Theorem 1]. A *ZPI-ring* is a general ZPI-ring in which every proper ideal can be *uniquely* expressed as a product of prime ideals. For integral domains a (general) ZPI-domain is the same as a Dedekind domain.

Proposition 2.1.3. *For an integral domain, the following notions coincide.*

(i) General ZPI-domain.
(ii) ZPI-domain.
(iii) Dedekind domain.

About the proof, for (i)⇔(iii) see [34, Theorem 37.8 (1)⇔(3)]; (ii)⇒(i) is obvious, and the implication (i)⇒(ii) follows easily from the following general lemma.

Lemma 2.1.4. ([34, Lemma 37.6]) *Let H be an invertible ideal in a commutative ring R. If H can be expressed as a finite product of proper prime ideals of R, then this representation is unique.*

E. Matlis, in 1964 [60, §8], called an integral domain R *h-local* if:

(a) Each nonzero ideal of R is contained in only finitely many maximal ideals (or, equivalently, R is a *domain with finite character*; i.e., each nonzero element of R belongs to a finite number of maximal ideals).
(b) Each nonzero prime is contained in a unique maximal ideal.

The notion of an h-local domain actually predates the work of Matlis. In 1952, P. Jaffard [52] considered the class of commutative rings such that each nonzero ideal can be factored as a finite product of ideals with each factor in a unique (and distinct) maximal ideal. He referred to a ring with this property as an *anneau du type de Dedekind*. It is clear that each nonzero prime ideal in such a ring is contained in a unique maximal ideal. Also, each nonzero nonunit is contained in only finitely many maximal ideals. Thus a domain of "Dedekind type" is h-local. The converse holds as well. The proof illustrates a technique that we will find useful in those cases where the domain in question satisfies condition (b) in the definition of h-local: each nonzero prime lies in a unique maximal ideal.

Theorem 2.1.5. (P. Jaffard [52, Théorème 6]) *Let R be a domain. Then R is h-local if and only if it is a domain of Dedekind type.*

Proof. If P is a nonzero prime with a factorization, $P = J_1 J_2 \cdots J_n$ with each J_i in a unique and distinct maximal ideal, then $n = 1$ and $P = J_1$. For a nonzero proper principal ideal rR, if $rR = I_1 I_2 \cdots I_n$ where each I_i is in a unique maximal ideal M_i, then clearly $M_1, M_2, \ldots, M_n$ are the only maximal ideals that contain r. Hence a domain of Dedekind type is h-local.

For the converse, we first show that if I is a nonzero ideal of R and M is a maximal ideal that contains I, then M is the only maximal ideal that contains $J := IR_M \bigcap R$. By way of contradiction, assume N is a maximal ideal other than M that contains J. Then each minimal prime of J that is contained in N is comaximal with M. Let Q be such a prime and let $q \in Q$ and $m \in M$ be such that $q + m = 1$. As Q is minimal over J, there is a positive integer n and an element $t \in R \backslash Q$ such that $tq^n \in J$. But, clearly, this puts $t \in JR_M \bigcap R = IR_M \bigcap R$, a contradiction. Hence M is the only maximal ideal that contains J.

Next, let $\{M_1, M_2, \ldots, M_n\}$ be the set of maximal ideals that contain I and set $I_i := IR_{M_i} \bigcap R$. Then M_i is the only maximal ideal that contains I_i. In particular, $I_i R_{M_j} = R_{M_j}$ for all $j \neq i$. Checking locally shows that $I = I_1 I_2 \cdots I_n$. □

The class of h-local domains simultaneously generalizes both Dedekind domains and local domains. Matlis used the notion of h-local domain in order to study integral domains R having (in common with Dedekind domains) the property that each torsion R-module $\mathscr{T}$ admits a primary decomposition (i.e., $\mathscr{T} \cong \bigoplus_{M \in \mathrm{Max}(R)} \mathscr{T}_M$ [60, Theorem 22]).

The h-local domain property was also used in the study of integral domains in which each nonzero ideal is divisorial. In 1968, Heinzer established the following characterization of integrally closed domains for which each nonzero ideal is divisorial.

Theorem 2.1.6. (Heinzer [44, Theorem 5.1]) *Let R be an integrally closed domain. Then each nonzero ideal of R is divisorial if and only if R is an h-local Prüfer domain such that each maximal ideal is invertible.*

Remark 2.1.7. With a different approach, Matlis, also in 1968, proved a related general result. Call a domain R *reflexive* if $\mathrm{Hom}_R(-, R)$ induces a duality on submodules of finite rank of free R-modules. *If R is a reflexive domain, then R is h-local* [61, Theorem 2.7] (see also [27, Proposition 5.6]). Note that a reflexive domain is always a *divisorial domain* (i.e., an integral domain such that every nonzero ideal is divisorial) and, for Noetherian domains, the two notions coincide [61] or [27, Propositions 5.6 and 5.8].

The following statements demonstrate more clearly why the h-local domain property is so useful when considering the local-global behavior of divisorial ideals.

Proposition 2.1.8. *Let R be an integral domain with quotient field K.*

(1) (Bazzoni-Salce [9, Lemma 2.3]) *If R is an h-local domain and $X \subseteq Y$ are R-submodules of K, then $(Y : X)R_M = (YR_M : XR_M)$ for each* $M \in \mathrm{Max}(R)$.
(2) (Olberding [67, Theorem 3.10]) *Let R be a Prüfer domain. The following statements are equivalent.*
   (i) *R is h-local.*
   (ii) $(Y : X)R_M = (YR_M : XR_M)$ *for all nonzero R-submodules X and Y of K such that $X \subseteq Y$ and for all* $M \in \mathrm{Max}(R)$.
   (iii) $(R : I)R_M = (R_M : IR_M)$ *for all nonzero ideals I of R and for all maximal ideals* $M \in \mathrm{Max}(R)$.

In a recent paper, Olberding [71] gives an ample survey of h-local domains, collecting several characterizations of different nature and discussing numerous examples of this distinguished class of integral domains.

Remark 2.1.9. Note that in studying Prüfer domains with Clifford class semigroup, S. Bazzoni [7, Theorem 2.14 and Corollary 3.4] proved: *If R is a Prüfer domain with finite character, then the following property holds:*

(loc-inv) *a nonzero ideal I of R is invertible if and only if IR_M is a nonzero principal ideal for every* $M \in \mathrm{Max}(R)$.

In that work [7, page 360], and in a following one [8, Question 6.2], she proposed the following conjecture: *A Prüfer domain R satisfies the property* **(loc-inv)** *if and only if R has finite character.* This conjecture was recently proved by W.C. Holland, J. Martinez, W.Wm. McGovern, and M. Tesemma in [49, Theorem 10] and, independently, by F. Halter-Koch in [41, Theorem 6.11]. For a purely ring-theoretic proof see M. Zafrullah [78] and W.Wm. McGovern [63].

We say that a maximal ideal M of an integral domain R is *unsteady* if MR_M is principal but M is not invertible. A maximal ideal is *steady* if it is not unsteady; i.e., if either MR_M is not principal or M is invertible. The following result is straightforward.

Lemma 2.1.10. *Let R be a Prüfer domain and $M \in \mathrm{Max}(R)$. Then*

(1) M is a steady maximal ideal of R if and only if it is either idempotent or invertible.
(2) If M is not invertible (or, equivalently, nondivisorial by Remark 2.1.2(1)), then either M is idempotent (i.e., steady) or M is locally principal (i.e., unsteady).

Note that from Remark 2.1.9, *unsteady maximal ideals do not exist in a Prüfer domain with finite character*. For a direct proof, note that if M is a maximal ideal of a Prüfer domain R (with finite character) such that MR_M is principal, then there is a nonzero element $r \in M$ such that $MR_M = rR_M$. By finite character, r is contained in at most finitely many other maximal ideals, say, $M_1, M_2, \ldots, M_n$. For each M_i, pick an element $t_i \in M \setminus M_i$. Then $B := t_1R + t_2R + \cdots + t_nR$ is contained in M but in none of the M_i. Checking locally, we have that $M = rR + B$, so M is invertible.

2.2 Sharp and Double Sharp Domains

An integral domain R is called a *#-domain* if

(#) for each pair of nonempty subsets Δ' and Δ'' of $\mathrm{Max}(R)$, $\Delta' \neq \Delta''$ implies $\bigcap\{R_{M'} \mid M' \in \Delta'\} \neq \bigcap\{R_{M''} \mid M'' \in \Delta''\}$ [34, page 331] (see also [32, page 817]).

This notion was introduced by Gilmer in [33]. The integral domains such that every overring is a #-domain were investigated in [35]; an integral domain with this property is called a *##-domain*.

Clearly, a Dedekind domain is a #-domain and, since an overring of a Dedekind domain is a Dedekind domain, a Dedekind domain is in fact a ##-domain [32, page 817 and Theorem 4(b)].

Recall that an integral domain R is an *almost Dedekind domain* if, for each maximal ideal M of R, R_M is a rank-one discrete valuation domain (or, equivalently, a (local) Dedekind domain). For instance, the integral closure of a rank-one discrete valuation domain (or, more generally, of an almost Dedekind domain) in a finite extension of its quotient field is an almost Dedekind domain [34, Theorem 36.1].

Even though a Dedekind domain is a ##-domain, an almost-Dedekind domain is not necessarily a #-domain [33, Sect. 3]. More precisely, in [33], one of the reasons Gilmer gives for considering the (#) property is as a way to distinguish between Dedekind domains and almost Dedekind domains that are not Dedekind. Specifically,

Theorem 2.2.1. (Gilmer [33, Theorem 3]) *An almost Dedekind domain is Dedekind if and only if it satisfies (#).*

It is easy to see [33, Lemma 1] that for an integral domain R we have:

$$R \textit{ is a \#-domain} \Leftrightarrow R_M \not\supseteq \bigcap\{R_N \mid N \in \mathrm{Max}(R), N \neq M\} \textit{ for all } M \in \mathrm{Max}(R).$$

For every ideal I of an integral domain R with quotient field K, set:

$$\begin{aligned} \mathrm{Max}(R,I) &:= \{N \in \mathrm{Max}(R) \mid N \supseteq I\}, \\ \Gamma_R(I) &:= \Gamma(I) := \bigcap\{R_M \mid M \in \mathrm{Max}(R,I)\} \quad \text{and} \\ \Theta_R(I) &:= \Theta(I) := \bigcap\{R_N \mid N \in \mathrm{Max}(R)\setminus\mathrm{Max}(R,I)\} \end{aligned}$$

(where we set $\Theta_R(I) := K$ if $\mathrm{Max}(R,I) = \mathrm{Max}(R)$ and, obviously, $\Theta_R(R) = R$). With this notation, the previous result can be stated as follows.

Proposition 2.2.2. (Gilmer [33, Lemma 1]) *Let R be an integral domain. Then R is a #-domain if and only if $R_M \not\supseteq \Theta(M)$ for all $M \in Max(R)$.*

Remark 2.2.3. With respect to Remark 2.1.9, Bazzoni in 1996 proved: *If R is a Prüfer #-domain, then R satisfies* **(loc-inv)** if and only if R is a domain with finite character [7, Theorem 4.3], providing a basis for her conjecture.

In the case of Prüfer domains, #-domains and ##-domains may be characterized as follows.

Theorem 2.2.4. (Gilmer–Heinzer [35, Theorems 1 and 3 and Corollary 2]) *Let R be a Prüfer domain.*

(1) The following statements are equivalent.

(i) R is a #-domain.
(ii) Each maximal ideal M of R contains a finitely generated ideal I such that $\mathrm{Max}(R,I) = \{M\}$.

(2) The following statements are equivalent.

(i) R is ##-domain.
(ii) Each prime ideal P of R contains a finitely generated ideal I such that $\mathrm{Max}(R,I) = \mathrm{Max}(R,P)$.
(iii) For each nonzero prime P of R, $R_P \not\supseteq \Theta(P)$.

The proofs of the statements collected in Theorem 2.2.4 can also be found in [24, Theorems 4.1.4, 4.1.6, and 4.17].

Condition (iii) of Theorem 2.2.4(2) has been adapted to considering individual primes in a general integral domain as being sharp or not. Specifically, a nonzero prime ideal of an integral domain R is called a *sharp prime* if $R_P \not\supseteq \Theta(P)$ [20] (cf. also [58, page 62]). Thus, by the previous considerations, *a Prüfer domain is a #-domain (respectively, a ##-domain) if and only if each nonzero maximal (respectively, prime) ideal is a sharp prime.*

Corollary 2.2.5. (Gilmer–Heinzer [35, Theorem 5]) *If R is a Prüfer domain with finite character, then R is a ##-domain and each nonzero prime is sharp.*

Proof. Let P be a nonzero prime of R and choose any nonzero element $r \in P$. By finite character, r is contained in at most finitely many maximal ideals that do not contain P. For each such maximal ideal M_i (if any), select an element $t_i \in P \setminus M_i$. Next set $B := rR + t_1R + t_2R + \cdots + t_nR$ where $M_1, M_2, \ldots, M_n$ are the maximal ideals that do not contain r. Clearly $\mathrm{Max}(R, B) = \mathrm{Max}(R, P)$. Thus R is ##-domain and each nonzero prime is sharp. □

2.3 Sharp and Antesharp Primes

J. Huckaba and I. Papick proved that if P is a nonzero prime ideal of a Prüfer domain R, then (in our notation) $\Theta(P) \bigcap R_P = (P : P) \subseteq P^{-1} \subseteq \Theta(P)$ [51, Theorems 3.2 and 3.8]. More precisely, we collect now some general results that we will need later and which imply the previous inclusions. We start with some further notation. For an ideal I of R, we let $\mathrm{Min}(R, I)$ denote the minimal primes of I (in R). Then we set:

$$\Phi_R(I) := \Phi(I) := \bigcap\{R_P \mid P \in \mathrm{Min}(R, I)\},$$
$$\Omega_R(I) := \Omega(I) := \bigcap\{R_P \mid P \in \mathrm{Spec}(R),\ P \not\supseteq I\}.$$

Obviously, $\Omega(I) \subseteq \Theta(I)$. Using this notation we have:

Lemma 2.3.1. (Hays [42, Lemma 1]) *If I is an ideal of an integral domain R, then $I^{-1} \subseteq \Omega(I)$ ($\subseteq \Theta(I)$).*

Theorem 2.3.2. (Huckaba–Papick [51, Theorems 3.2 and 3.8, Lemma 3.3, Proposition 3.9]; Fontana–Huckaba–Papick [21]) *Let R be a Prüfer domain.*

(1) Let I be a nonzero ideal of R. Then the following hold.

(a) $\Theta(I) \bigcap \Phi(I) \subseteq I^{-1} \subseteq \Theta(I)$.
(b) Moreover, I^{-1} is a ring implies $\Theta(I) \bigcap \Phi(I) = I^{-1}$.

(2) Let P be a nonzero prime ideal of R. Then the following hold.

(a) $\Theta(P) \bigcap R_P \subseteq P^{-1} \subseteq \Theta(P)$.
(b) Moreover, P noninvertible implies $(P : P) = P^{-1} = \Theta(P) \bigcap R_P$. In particular, $PP^{-1} = P$.

(3) $Q^{-1} = (Q : Q)$ for each nonzero nonmaximal prime Q of R.

The proofs of the statements in Theorem 2.3.2 can also be found in [24, Theorem 3.1.2, Corollary 3.1.8 and Lemma 4.1.9]. (Note that the inclusion $I^{-1} \subseteq \Theta(I)$ in (1)(a) is trivial consequence of the previous Lemma 2.3.1 since in general $\Omega(I) \subseteq \Theta(I)$.)

Remark 2.3.3. With respect to Theorem 2.3.2(2)(b), note that the equality $P^{-1} = (P : P)$ holds in more general settings. For instance, in [59, Lemma 15], Lucas proved: *Given a nonzero ideal I in an integrally closed domain, I^{-1} is a ring if and only if $I^{-1} = (\sqrt{I})^{-1} = (\sqrt{I} : \sqrt{I})$.* An even more general result is given in [24, Proposition 3.1.16].

Lemma 2.3.4. (Gilmer–Heinzer [35, Corollaries 2 and 3]) *Let R be a Prüfer domain.*

(1) Let P be a prime ideal of R and let $\{M_\alpha\}$ be the set of maximal ideals of R not containing P. Then $R_P \not\supseteq \bigcap_\alpha R_{M_\alpha}$ if and only if there exists a finitely generated ideal I of R contained in P such that I is contained in no M_α.
(2) Let $M \in \mathrm{Max}(R)$. If there exists $m \in M$ such that m belongs to only finitely many maximal ideals of R, then $R_M \not\supseteq \bigcap\{R_N \mid N \in \mathrm{Max}(R),\ N \neq M\}$. Hence, if for each $M_\alpha \in \mathrm{Max}(R)$ there is an element $m_\alpha \in M_\alpha$ such that m_α belongs to only finitely many maximal ideals of R, then R is a #-domain.

From Theorem 2.3.2 and Lemma 2.3.4(1), we have the following straightforward consequences.

Proposition 2.3.5. *Let R be a Prüfer domain and let P be a nonzero prime ideal of R.*

(1) The following statements are equivalent.

(i) P is a sharp prime (i.e., $R_P \not\supseteq \Theta(P)$).
(ii) $\mathrm{Max}(R, I) = \mathrm{Max}(R, P)$ for some finitely generated ideal $I \subseteq P$.
(iii) $P^{-1} \subsetneq \Theta(P)$.

(2) Assume also that P is noninvertible. Then the previous statements are equivalent to the following.

(iv) $(P : P) \subsetneq \Theta(P)$.

Remark 2.3.6. Note that Gilmer and Heinzer [36, Proposition 1.4] a more general version of Lemma 2.3.4(1): *Let $\{P_\alpha\} \bigcup \{P\}$ be a set of primes in a Prüfer domain R, then $\bigcap_\alpha R_{P_\alpha} \subseteq R_P$ if and only if each finitely generated ideal contained in P is contained in some P_α.*

For a nonzero nonmaximal prime ideal P of a Prüfer domain R, we know that P is divisorial if and only if $P^{-1} \subsetneq \Theta(P)$ or $(R : \Theta(P)) = P$ [24, Theorem 4.1.10]. Therefore, under the present assumptions on P, we have that P is sharp implies that P is divisorial [21, Theorem 2.1]. In particular, for a Prüfer ##-domain, we are immediately able to determine some types of divisorial ideals.

Proposition 2.3.7. (Fontana–Huckaba–Papick [21, Corollary 2.6], [22, Proposition 12]) *Let R be a Prüfer domain which is also a ##-domain.*

(1) If P is a nonzero nonmaximal prime ideal of R, then P is divisorial.
(2) The product of divisorial prime ideals is divisorial. In particular, for each nonzero nonmaximal prime ideal P of R, P^e is divisorial for all integers $e \geq 1$.

The proof of this proposition can also be found in [24, Theorem 4.1.21].

Remark 2.3.8. With respect to Proposition 2.3.7(2), note that a prime ideal P of a Prüfer domain can be divisorial, even if P^2 is not [24, Example 8.4.1]. On the other hand, it is known that $P^2 = (P^2)^v$ if and only if $P^e = (P^e)^v$ for all integers $e \geq 1$ [24, Corollary 4.1.18 and Theorem 4.1.19].

For general integral domains, sharp primes have a sort of "stability property under specializations." More precisely,

Proposition 2.3.9. *Let $P \subseteq P'$ be a pair of nonzero prime ideals of a domain R. If P is sharp and $\mathrm{Max}(R, P) = \mathrm{Max}(R, P')$, then P' is sharp. In particular, if P is sharp and contained in a unique maximal ideal M of R, then each prime that contains P is sharp.*

Proof. Since P' contains P, $R_{P'} \subseteq R_P$. If P is sharp and $\mathrm{Max}(R, P) = \mathrm{Max}(R, P')$, then R_P does not contain $\Theta(P)$ and $\Theta(P) = \Theta(P')$. Hence we also have that $R_{P'}$ does not contain $\Theta(P')$. Therefore P' is sharp. The "in particular" statement follows easily. □

Recall that a *branched prime* is a prime ideal P having a proper P-primary ideal. An *unbranched prime* is a prime ideal that is not branched. Sharp prime ideals that are also branched can be easily characterized in Prüfer domains.

Proposition 2.3.10. *Let R be a Prüfer domain and let P be a nonzero prime ideal of R. Then P is both sharp and branched if and only if $P = \sqrt{B}$ for some finitely generated ideal B of R.*

Proof. Since R is a Prüfer domain, P is branched if and only it is minimal over some finitely generated ideal [34, Theorem 23.3(e)]. Hence, if P is the radical of a finitely generated ideal, then it is both branched and sharp, the latter conclusion by Proposition 2.3.5(1). Conversely, if P is both sharp and branched, then it is minimal over a finitely generated ideal J and it contains a finitely generated ideal I such that the only maximal ideals that contain I are those that contain P. Clearly, P is the radical of the finitely generated ideal $I + J$ in this case. □

Regarding the proof of Proposition 2.3.10, note that if Q is a prime ideal that contains I, then Q is contained in some maximal ideal that contains I, and thus in a maximal ideal that contains P. Since R is a Prüfer domain, P and Q must be comparable.

The next goal is to provide further characterizations for general sharp primes of a Prüfer domain.

Theorem 2.3.11. *Let R be a Prüfer domain and let P be a nonzero prime ideal of R. The following statements are equivalent.*

(i) *P is sharp.*
(ii) *There is a prime ideal $Q \subseteq P$ such that Q is the radical of a finitely generated ideal with $\Theta(Q) = \Theta(P)$.*

(iii) *There is a prime ideal $Q \subseteq P$ such that Q is the radical of a finitely generated ideal and each maximal ideal that contains Q also contains P.*
(iv) *There is a prime ideal $Q \subseteq P$ such that Q is the radical of a finitely generated ideal and each prime that contains Q is comparable with P.*
(v) *There is a prime ideal $Q \subseteq P$ such that Q is the radical of a finitely generated ideal and each ideal that contains Q is comparable with P.*
(vi) *There is a finitely generated ideal $I \subseteq P$ such that each ideal that contains I is comparable with P.*

Moreover, the prime ideal Q in statements (ii)– (v) is sharp and branched.

Proof. Clearly (v) implies (iv), (iv) implies (iii), and (iii) implies (ii). Moreover, (ii) implies (i) by Propositions 2.3.5 and 2.3.10.

We show next that (i) implies (v) and (vi).

Assume P is sharp and let $I \subseteq P$ be a finitely generated ideal such that each maximal ideal that contains I also contains P (Proposition 2.3.5(1)). Let Q be a prime minimal over I. Then each maximal ideal M that contains Q also contains P. Since R_M is a valuation domain (or, since Spec(R) is treed), P must contain Q. It follows that Q is the unique minimal prime of I, and therefore $Q = \sqrt{I}$.

For the rest, it suffices to start with an element $x \in R \setminus P$ such that $xR + P \neq R$, and then show that the ideal $J := xR + I$ contains P. For this, we simply see what happens when we localize at a maximal ideal. Clearly, if N is a maximal ideal that does not contain I, then $JR_N = R_N = PR_N$. On the other hand, if M is a maximal ideal that contains I, then it also contains P. Since $x \notin P$ and so $x \notin I$, we have $JR_M = xR_M + IR_M = xR_M \supsetneq PR_M$. It follows that $J \supsetneq P$.

To finish we show that (vi) implies (v). Assume I is a finitely generated ideal of R with $I \subseteq P$ and each ideal containing I is comparable with P. Let Q be a prime minimal over I. Since Q is comparable with P, we must have $Q \subseteq P$. Since Spec(R) is treed, Q must be the unique minimal prime of I, so that $Q = \sqrt{I}$.

The last statement is a consequence of Proposition 2.3.10. □

We collect in the next statement several characterizations of branched sharp primes of a Prüfer domain.

Theorem 2.3.12. *Let P be a nonzero prime of a Prüfer domain R.*

(1) The following statements are equivalent.

(i) *P is sharp and branched.*
(ii) *P is the radical of a finitely generated ideal.*
(iii) *$P\Omega(P) = \Omega(P)$.*
(iv) *$\Omega(P) \nsubseteq R_P$.*
(v) *$P^{-1} \subsetneq \Omega(P)$.*
(vi) *If Q is a proper P-primary ideal, then*
$QQ^{-1} = P$, *whenever P is not maximal and*
$QQ^{-1} \supseteq P$, *whenever P is maximal.*
(vii) *There exists a proper P-primary ideal that is divisorial.*

(2) Assume in addition that P is not a maximal ideal of R. Then the previous statements are equivalent to the following.

(viii) Each P-primary ideal of R is divisorial.

The proof of this result can be found in [20, Propositions 2.7 and 2.9]. (Note that the equivalence of (i) with (ii) was already proved in Proposition 2.3.10 and the equivalence of (iv) with (v) follows easily from the following inclusions (Theorem 2.3.2(2) and Lemma 2.3.1):

$$\Omega(P)\bigcap R_P \subseteq \Theta(P)\bigcap R_P \subseteq P^{-1} \subseteq \Omega(P).)$$

Before developing alternate characterizations of sharp primes, we present a few basic results dealing with duals of ideals, trace ideals and endomorphism rings.

Recall from above that for a nonzero ideal I of R, $\Gamma(I)$ is the intersection of the localizations R_M at the maximal ideals M that contain I and $\Theta(I)$ is the intersection of the localizations R_N at the maximal ideals N that do not contain I. Hence it is clear that $\Gamma(I)\bigcap\Theta(I) = R$. Also note that $\Gamma(I) = \Gamma(J)$ and $\Theta(I) = \Theta(J)$, for all ideals J with $I \subseteq J \subseteq \bigcap\{M \mid M \in \mathrm{Max}(R,I)\}$; in particular, $\Gamma(I) = \Gamma(\sqrt{I})$ and $\Theta(I) = \Theta(\sqrt{I})$. Moreover, for each ideal B of R, $B \subseteq B\Gamma(I) \subseteq BR_M$ for each $M \in \mathrm{Max}(R,I)$ and $B \subseteq B\Theta(I) \subseteq BR_N$ for each $N \in \mathrm{Max}(R)\backslash\mathrm{Max}(R,I)$ (if any). Hence, for each ideal B of R, $B = B\Gamma(I)\bigcap B\Theta(I)$. Recall also that an ideal I of a domain R is a *trace ideal* if $I = II^{-1}$ (equivalently, $I^{-1} = (I : I)$) (see Sect. 2.4).

Lemma 2.3.13. *Let I and J be a nonzero ideals of an integral domain R. If S and T are overrings of R such that $S\bigcap T = R$, then $J^{-1} = (S : JS)\bigcap(T : JT)$. In particular, $J^{-1} = (\Gamma(I) : J)\bigcap(\Theta(I) : J)$, and $I^{-1} = (\Gamma(I) : I)\bigcap\Theta(I)$.*

Proof. Assume S and T are overrings of R such that $S\bigcap T = R$. For J, it is clear that both $(S : J) = (S : JS)$ and $(T : J) = (T : JT)$ contain $J^{-1} = (R : J)$. For the reverse containment, if $s \in J$ and $t \in (S : J)\bigcap(T : J)$, then $st \in S\bigcap T = R$. Thus $J^{-1} = (S : JS)\bigcap(T : JT)$.

It is always the case that $\Gamma(I)\bigcap\Theta(I) = R$ when I is a nonzero ideal of R. Thus $J^{-1} = (\Gamma(I) : J)\bigcap(\Theta(I) : J)$.

For $N \in \mathrm{Max}(R)\backslash\mathrm{Max}(R,I)$, $I^{-1} \subseteq R_N$. Thus $I^{-1} \subseteq \Theta(I) \subseteq (\Theta(I) : I)$ and $I^{-1} = (\Gamma(I) : I)\bigcap\Theta(I)$. □

Lemma 2.3.14. *Let J be a nonzero trace ideal of an integral domain R, and let P be a prime that contains J. Then $B := JR_P\bigcap R$ is also a trace ideal of R.*

Proof. Since $J \subseteq B$, $B^{-1} \subseteq J^{-1}$. On the other hand, $J = JJ^{-1}$ and $BR_P = JR_P$, whence $BB^{-1}R_P \subseteq BJ^{-1}R_P = JJ^{-1}R_P = JR_P = BR_P$. It follows that $BB^{-1} = B$. □

The proof of the next lemma requires us to look ahead to Sect. 2.4. Specifically, we need to know that if Q is a nonzero primary ideal of a valuation domain such that $\sqrt{Q}$ is not the maximal ideal, then $QQ^{-1} = \sqrt{Q}$ (see Propositions 2.4.1 and 2.4.9).

Lemma 2.3.15. *Let R be a Prüfer domain, and let Q be a nonzero primary ideal of R with radical P.*

(1) $(Q : Q) = (P : P) = R_P \bigcap \Theta(P)$.
(2) If P is maximal, then $Q^{-1} = (R_P : Q) \bigcap \Theta(P)$.
(3) If P is not maximal, then $QQ^{-1} \subseteq P$.
(4) If $QQ^{-1} \subseteq P$, then $Q^{-1} = (PR_P : Q) \bigcap \Theta(P)$. In particular, $Q^{-1} = (PR_P : Q) \bigcap \Theta(P)$ whenever P is not maximal.

Proof. First note that since $\sqrt{Q} = P$, $\Gamma(Q) = \Gamma(P)$ and $\Theta(Q) = \Theta(P)$.

For (1), PR_M is the prime ideal of all zero divisors on both R_M/QR_M and R_M/PR_M for each maximal ideal $M \in \mathrm{Max}(R, P)$. Hence $(Q : Q) = (P : P) = R_P \bigcap \Theta(P)$ by Theorem 2.3.2(2) and Remark 2.1.2(4).

Statement (2) is simply a restatement of the last conclusion in Lemma 2.3.13 since $\Gamma(Q) = R_P$ when P is maximal (and $\Theta(Q) = \Theta(P)$).

For (3), if P is not maximal and M is a maximal ideal that contains P, then $QQ^{-1}R_M \subseteq (QR_M)(QR_M)^{-1} = PR_M$ since R_M is a valuation domain (as per the observation above). Hence $QQ^{-1} \subseteq P$.

For (4) assume $QQ^{-1} \subseteq P$. Then Q^{-1} is also contained in $(PR_P : Q)$. Since it is always the case that $Q^{-1} \subseteq \Theta(P)$, $Q^{-1} \subseteq (PR_P : Q) \bigcap \Theta(P)$. For the reverse containment, suppose $t \in (PR_P : Q) \bigcap \Theta(P)$ and $q \in Q$. Then $tq \in PR_P \bigcap \Theta(P)$. Hence, for any maximal ideal $M \supseteq P$, we have $tq \in PR_P = PR_M \subseteq R_M$, and we have $tq \in PR_P \bigcap \Gamma(P) \bigcap \Theta(P) = P$. Thus $Q^{-1} = (PR_P : Q) \bigcap \Theta(P)$. □

The first two statements in the next result are easily derived from Theorem 2.3.12(1). The third statement is simply a combination of Theorem 2.3.2(2) and Lemma 2.3.15(1) (or Remark 2.1.2(4)).

Lemma 2.3.16. *Let Q be a proper P-primary ideal of a Prüfer domain R.*

(1) If Q^{-1} is not a ring, then either $QQ^{-1} = P$ or both $QQ^{-1} = R$ and P is maximal.
(2) If QR_P is invertible and P is both sharp and maximal, then Q is invertible.
(3) If Q^{-1} is a ring, then $Q^{-1} = (Q : Q) = (P : P) = P^{-1}$.

Proof. For (1), assume Q^{-1} is not a ring. Since $\Theta(P) = \Theta(Q)$ contains Q^{-1} (Lemma 2.3.13), we must have proper containment. Thus $\Theta(P) \supsetneq Q^{-1} \supsetneq (Q : Q) = R_P \bigcap \Theta(P)$ and we have that P is sharp by definition. Hence, by Theorem 2.3.12(1), $QQ^{-1} = P$ whenever P is not maximal, and $QQ^{-1} \supseteq P$ whenever P is maximal. This gives (1).

For (2), assume QR_P is invertible and P is both sharp and maximal. Then $QR_P = tR_P$ for some $t \in Q$. Also, by Proposition 2.3.10, $P = \sqrt{B}$ for some finitely generated ideal B. Since $Q \supseteq B^m$ for some positive integer m, checking locally shows that $Q = tR + B^m$. Thus Q is invertible.

Finally for (3), the fact that $(Q : Q) = (P : P) = R_P \bigcap \Theta(P) \subseteq P^{-1} \subseteq Q^{-1}$ follows from Lemma 2.3.15(1) whether Q^{-1} is a ring or not. If Q^{-1} is a ring, then we have $Q^{-1} = (Q : Q) = (P : P) = P^{-1}$ by Theorem 2.3.2 and Lemma 2.3.15(1). □

Our next result adds more ways of characterizing sharp branched primes.

Theorem 2.3.17. *The following are equivalent for a nonzero branched prime P of a Prüfer domain R.*

(i) P is sharp.
(ii) $Q^{-1}R_M = (QR_M)^{-1}$ for each P-primary ideal Q and each maximal ideal M of R.
(iii) There is a proper P-primary ideal Q and a maximal ideal $M \in \mathrm{Max}(R, P)$ such that $Q^{-1}R_M = (QR_M)^{-1}$.
(iv) There is a P-primary ideal Q such that Q^{-1} is not a ring.
(v) For each proper P-primary ideal Q of R, Q^{-1} is not a ring.
(vi) If P is a minimal over a trace ideal I of R, then $IR_M = PR_M$ for each maximal ideal $M \in \mathrm{Max}(R, P)$.

Proof. With regard to statements (ii) and (iii), if N is a maximal ideal that does not contain P, then $QR_N = R_N$ and $Q^{-1} \subseteq R_N$ for each P-primary ideal Q. Hence $Q^{-1}R_N = (QR_N)^{-1} = R_N$ in this case, regardless of whether P is sharp or not.

Obviously, (ii) implies (iii), and (v) implies (iv).

For (i) implies (ii), assume P is sharp and let Q be a P-primary ideal. As the valuation domain R_P does not contain $\Theta(P)$, $\Theta(P)R_P = R_{P_0}$ for some prime $P_0 \subsetneq P$ (possibly with $P_0 = (0)$) and $QR_PR_{P_0} = R_{P_0}$ which puts $(R_P : Q)$ inside $R_{P_0} = \Theta(P)R_P$. First, suppose that P is maximal. If QR_P is invertible, then Q is invertible by Lemma 2.3.16(2), and in this case $Q^{-1}R_M = (QR_M)^{-1}$ for each maximal ideal M. On the other hand, if QR_P is not invertible, then $QR_P(QR_P)^{-1} = PR_P$ since the valuation domain R_P has the trace property (Proposition 2.4.1). Hence $(QR_P)^{-1} = (PR_P : QR_P) = (PR_P : Q)$. By Lemma 2.3.15(4), $Q^{-1} = (PR_P : Q) \bigcap \Theta(P)$, whence $Q^{-1}R_P = ((PR_P : Q) \bigcap \Theta(P))R_P = (PR_P : Q) \bigcap \Theta(P)R_P = (PR_P : Q) = (QR_P)^{-1}$.

For the case where P is not maximal, let M be a maximal ideal that contains P. Then $QR_M(QR_M)^{-1} = PR_M = PR_P$ by Propositions 2.4.1 and 2.4.9. Since P is sharp, $P\Theta(P) = \Theta(P)$. Thus $\Theta(P)R_M$ is a proper overring of R_M in which P blows up and is therefore a proper overring of R_P. Hence, by Lemma 2.3.15(4), $Q^{-1}R_M = ((PR_P : Q) \bigcap \Theta(P))R_M = (PR_P : Q)R_M \bigcap \Theta(P)R_M = (PR_P : Q)R_M = (PR_M : QR_M) = (R_M : QR_M) = (QR_M)^{-1}$. Therefore (i) implies (ii).

To see that (iii) implies (iv), assume there is a proper P-primary ideal Q and a maximal ideal $M \in \mathrm{Max}(R, P)$ such that $Q^{-1}R_M = (QR_M)^{-1}$. In a valuation domain, each proper primary ideal has a nontrivial inverse that cannot be a ring (Lemma 2.1.1(2)). Thus Q^{-1} cannot be a ring. The same reasoning proves that (ii) implies (v).

For (iv) implies (i), let Q be a P-primary ideal such that Q^{-1} is not a ring. Since $(P : P) = R_P \bigcap \Theta(P) \subseteq P^{-1} \subseteq Q^{-1} \subseteq \Theta(Q) = \Theta(P)$ in all cases (by Theorem 2.3.2(2)), having Q^{-1} not a ring implies $R_P \bigcap \Theta(P)$ is properly contained in $\Theta(P)$. Hence P is sharp.

Assume the statement in (vi) holds, and let Q be a proper P-primary ideal. If Q^{-1} is a ring, then Q is a trace ideal by Lemma 2.3.16(3). But, by assumption, this implies $QR_M = PR_M$ for each maximal ideal $M \in \text{Max}(R, P)$, a contradiction. Thus it must be that Q^{-1} is not a ring. Hence (vi) implies (v).

To finish the proof, we show that (i) implies (vi). Assume P is sharp and let I be a trace ideal with P minimal over I. Then P is a trace ideal by [48, Proposition 2.1] (or see the proof of ((ii)⇒(i)) for Corollary 2.4.13 and apply Theorem 2.3.2(2)). Hence, by (ii), $R_P = P^{-1}R_M = (PR_M)^{-1}$ for each $M \in \text{Max}(R, P)$. By Lemma 2.3.14, $IR_P \bigcap R$ is a P-primary ideal that is also a trace ideal. Since P is sharp, we have $IR_P \bigcap R = P$ by Theorem 2.3.12(1). Thus $PR_M = PR_P = IR_P = IP^{-1}R_M \subseteq II^{-1}R_M = IR_M$. Therefore $IR_M = PR_M$ for each $M \in \text{Max}(R, P)$, and the proof is complete. □

Using Theorem 2.3.17 and Lemmas 2.3.15 and 2.3.16, we can give a characterization of branched primes ideals that are not sharp.

Corollary 2.3.18. *The following statements are equivalent for a nonzero branched prime ideal P of a Prüfer domain R.*

(i) *P is not sharp.*
(ii) $(P : P) = \Theta(P)$.
(iii) $Q^{-1} = \Theta(P)$ *for each P-primary ideal Q.*
(iv) $Q^{-1} = \Theta(P)$ *for some proper P-primary ideal Q.*

Proof. Proper P-primary ideals exist since P is branched. By Theorem 2.3.17[(i) ⇒(v)], if P is sharp, then there is no proper P-primary ideal Q such that Q^{-1} is a ring. Thus (iii) implies both (i) and (iv).

Let Q be a P-primary ideal. Then $\Theta(P) \bigcap R_P \subseteq P^{-1} \subseteq Q^{-1} \subseteq \Theta(P)$ by Theorem 2.3.2(1)(a). Also, by Lemma 2.3.15, $(Q : Q) = (P : P) = \Theta(P) \bigcap R_P$. If $(P : P) = \Theta(P)$, then we have $Q^{-1} = \Theta(P)$. Moreover, if $Q^{-1} = \Theta(P)$, then we have $\Theta(P) = Q^{-1} = (Q : Q) = (P : P)$ (Lemma 2.3.16). Thus (ii), (iii) and (iv) are equivalent. If P is not sharp, then $R_P \supseteq \Theta(P)$ and we then have $(P : P) = \Theta(P)$. Hence (i) implies (ii). □

For an unbranched maximal ideal M of a Prüfer domain R, M is sharp if and only if there is an infinite chain of sharp branched primes $\{P_\alpha\}$ with $\bigcup P_\alpha = M$ and $\text{Max}(R, P_\alpha) = \{M\}$ [20, Proposition 2.10]. A similar condition characterizes nonmaximal unbranched sharp primes, as we now show.

Theorem 2.3.19. *Let P be a nonzero unbranched prime of a Prüfer domain R. Then P is sharp if and only if there is an infinite chain of sharp branched primes $\{P_\alpha\}$ such that $\bigcup P_\alpha = P$ with* $\text{Max}(R, P) = \text{Max}(R, P_\alpha)$ *for each P_α.*

Proof. Assume P is sharp. Then it is both sharp and maximal in the ring $(P : P)$ (Corollary 2.3.21). For each prime $Q \subseteq P$, $Q(P : P) = Q$. Thus, by

[20, Proposition 2.10], there is an infinite chain of branched primes $\{P_\alpha\}$ in $(P : P)$ (and in R) with $\bigcup P_\alpha = P$ and $\text{Max}((P : P), P_\alpha) = \{P\}$. Since each maximal ideal of R that does not contain P survives in $(P : P)$, $\text{Max}(R, P_\alpha) = \text{Max}(R, P)$ for each P_α.

For the converse, if P' is a sharp branched prime that is contained in P such that $\text{Max}(R, P) = \text{Max}(R, P')$, then P is sharp by Lemma 2.3.9. □

A nonzero prime P of an integral domain R is said to be *antesharp* if each maximal ideal of $(P : P)$ that contains P, contracts to P in R [20, Sect. 1].

We recall in the next proposition some characterizations of antesharp prime ideals, paying particular attention to the Prüfer domain case [20, Proposition 2.3].

Proposition 2.3.20. *Let P be a nonzero nonmaximal prime ideal of an integral domain R. Then the following are equivalent.*

(i) P is antesharp.
(ii) For each $a \in R \setminus P$, the ideal $A := aR + P$ is invertible.
(iii) For each prime Q of R that properly contains P, there is an invertible ideal $I \subseteq Q$ that properly contains P.

If, in addition, R is Prüfer, then (i), (ii) and (iii) are also equivalent to the following.

(iv) P is a maximal ideal of $(P : P)$
(v) Each prime ideal of $(P : P)$ that contains P contracts to P in R and is a maximal ideal of $(P : P)$.
(vi) For each prime Q of R that properly contains P, there is a finitely generated ideal $I \subseteq Q$ that properly contains P.

If R is not a Prüfer domain, then it is the still the case that (i) implies (vi). However, the reverse implication does not hold in general: a simple example is the prime $P = (X, Y)$ of the polynomial ring $K[X, Y, Z]$ where K is a field. Also note that since each invertible ideal of a local domain is principal, if R is local, then P is antesharp if and only if it is divided; i.e., P compares with each principal ideal.

From Proposition 2.3.20, we deduce that a sharp prime of a Prüfer domain is always antesharp, but a (nonmaximal) antesharp prime need not be sharp (see for example [20, Example 4.9]). More precisely [20, Corollary 2.4],

Corollary 2.3.21. *Let P be a (nonzero) prime ideal of an integral domain R.*

(1) If P is antesharp and not maximal, then it is divisorial.
(2) If, moreover, R is a Prüfer domain and P is sharp, then P is a sharp maximal ideal of $(P : P)$ and antesharp as a prime of R. Furthermore, if P is sharp and not maximal, then it is both antesharp and divisorial.

We give next a new characterization of nonmaximal antesharp ideals in a Prüfer domain.

Theorem 2.3.22. *Let P be a nonzero prime of a Prüfer domain R.*

(1) If P is an idempotent maximal ideal, then $P^{-1}R_N = (PR_N)^{-1}$ for each maximal ideal N.

(2) If P is not maximal, then P is antesharp if and only if $P^{-1}R_M = (PR_M)^{-1}$ for each maximal ideal M of R.

Proof. For (1), if P is an idempotent maximal ideal of a Prüfer domain R, then $P^{-1} = R$ and $(PR_P)^{-1} = R_P$ (Remark 2.1.2(1)). As in the proof of Theorem 2.3.17, it is also the case that $P^{-1}R_N = (PR_N)^{-1} = R_N$ for each maximal ideal $N \neq P$.

For (2), assume P is not maximal. As above, if N is a maximal ideal that does not contain P, then $PR_N = R_N = P^{-1}R_N = (PR_N)^{-1}$. Thus for (2), we only need to consider what happens with regard to those maximal ideals M that contain P.

Since P is not maximal, $P^{-1} = (P : P) = R_P \bigcap \Theta(P)$ (Theorem 2.3.2(2)), regardless of whether P is antesharp or not (Corollary 2.3.21). Also, for each maximal ideal M of R that contains P, considering the valuation domain R_M, it is well known that $(PR_M)^{-1} = (PR_M : PR_M) = R_P$ (Lemma 2.1.1).

Assume P is not antesharp. Then P is not sharp (Corollary 2.3.21(2)) and there is a prime P' that properly contains P and survives in $(P : P)$. Hence $P^{-1} = (P : P) = \Theta(P) \subseteq R_{P'}$. Let M be a maximal ideal that contains P'. Then $P^{-1}R_M = \Theta(P)R_M \subseteq R_{P'} \subsetneq R_P = (PR_M)^{-1}$.

For the converse, assume P is antesharp. Then P is a maximal ideal of $(P : P)$. Let M be a maximal ideal that contains P and let P_0 be the largest prime that is contained in M and survives in $\Theta(P)$ (use Zorn's Lemma). Since $\Theta(P)$ contains $(P : P)$ and P is maximal in $(P : P)$, P_0 is contained in P. Thus $P^{-1}R_M = (R_P \bigcap \Theta(P))R_M = R_P \bigcap R_{P_0} = R_P = (PR_M)^{-1}$. □

2.4 Trace Properties

Recall that given a ring R and an R-module B, the *trace of* B is the ideal of R generated by the set $\{r \in R \mid r = f(b)$ for some $b \in B$ and $f \in \mathrm{Hom}_R(B, R)\}$ (see for example, [24, Sect. 4.2]).

As we are dealing exclusively with integral domains R, $\mathrm{Hom}_R(I, R)$ is naturally isomorphic to I^{-1} ($:= (R : I)$) for each nonzero ideal I of R, and the trace of I is simply the ideal II^{-1}. More generally, given an integral domain R, if C is the trace of an R-module B, then $C^{-1} = (C : C)$ and so $C = CC^{-1}$ [24, Lemma 4.2.3].

As in [59] (and above in Sect. 2.3), we say that a (nonzero) ideal I of a domain R is a *trace ideal* of R if $II^{-1} = I$, or equivalently, if $I^{-1} = (I : I)$. An integral domain R has the *trace property* (or is a *TP-domain*) if each (proper) trace ideal is prime [23, page 169] (equivalently, II^{-1} is prime for each nonzero noninvertible ideal I of R), and it has the *radical trace property* (or is an *RTP-domain*) if each trace ideal is a radical ideal [46, page 110] (equivalently, II^{-1} is a radical ideal for each nonzero noninvertible ideal I). *If R is a Noetherian domain,* *then R is a RTP-domain if and only if R_P is a TP-domain for each $P \in Spec(R)$ [46, Proposition 2.1].*

Proposition 2.4.1. (D.D. Anderson–Huckaba–Papick [2, Theorem 2.8] and Fontana–Huckaba–Papick [23, Proposition 2.1]) *Every valuation domain is a TP-domain.*

A proof of this result can also be found in [24, Proposition 4.2.1]. Note that a local one-dimensional integrally closed domain with the trace property is not necessarily a valuation domain: an example is given by the domain $k + Yk(X)[Y]_{(Y)}$, where k is a field and X and Y are two indeterminates over k [24, Example 8.4.4].

Clearly, *every Dedekind domain is a TP-domain* [23, Corollary 2.5]. But not every Prüfer domain has the trace property. In fact, there is an example of an almost Dedekind domain with a unique noninvertible maximal ideal M such that $M^2(M^2)^{-1}$ is not a prime ideal (cf. [38, Sect. 6], [23, Example 4.3], and [24, Example 4.2.10]). On the other hand, it is known [23, Propositions 2.8 and 2.9] that the following properties hold:

coherent integrally closed TP-domain $\Rightarrow$ *Prüfer domain,*
Noetherian integrally closed TP-domain $\Leftrightarrow$ *Dedekind domain.*

Proposition 2.4.2. (Fontana–Huckaba–Papick [23, Proposition 2.10, Corollary 2.11]) *Let R be a TP-domain. If M is a noninvertible maximal ideal of R, then each noninvertible ideal of R is contained in M. In particular, if R has a noninvertible maximal ideal, then all the other maximal ideals of R are invertible.*

For one proof of Proposition 2.4.2, see also [24, Lemma 2.4.7, Corollary 2.4.8].

Even more can be said about the ideals of a TP-domain. The following four lemmas are quite useful in dealing with the various trace properties and with factorizations.

Lemma 2.4.3. (Lucas [59, Lemmas 0 and 1]) *Let P be a nonzero prime ideal of a domain R.*

(1) If PP^{-1} properly contains P and I is an ideal such that $P \subsetneq I \subseteq PP^{-1}$, then $(R : I) = (P : P) = (R : PP^{-1}) = (PP^{-1} : PP^{-1})$.
(2) If B is a radical ideal contained in P, but P is not minimal over B, then $(R : P) \subseteq (B : B)$.

Proof. For (1), first note that $(P : P) \subseteq (PP^{-1} : PP^{-1}) = (R : PP^{-1}) \subseteq (R : I)$. To establish the reverse containment, let $t \in (R : I)$. As both tI and tP are contained in R, $(tP)I = (tI)P \subseteq P$ implies $tP \subseteq P$. Therefore $(R : I) = (P : P)$.

For statement (2), let $q \in (R : P)$. Then $qP \subseteq R$ and $qB \subseteq R$. Hence $qPB \subseteq B$. As P is not minimal over B, qB is contained in each minimal prime of B. Thus $qB \subseteq B$. □

Lemma 2.4.4. *Let P be a nonzero prime ideal of a domain R. If $PP^{-1} \supsetneq P$ and for each $b \in PP^{-1}\backslash P$ there is a pair of elements $s \in (P : P)$ and $p \in P$ such that $b = b^2s + p$, then P is an invertible maximal ideal of R.*

Proof. Suppose $PP^{-1} \supsetneq P$. We have $at^n \in PP^{-1}$ for each $a \in PP^{-1}\backslash P$ and $t \in (P : P)$ since $(P : P) = (PP^{-1} : PP^{-1}) = (PP^{-1})^{-1}$ (Lemma 2.4.3). Further assume that for each $b \in PP^{-1}\backslash P$, there is a pair of elements $s \in (P : P)$ and $p \in P$ such that $b = b^2s + p$. Then $b(1 - bs) = p \in P$ with $1 - bs \in R$. Since P is prime, $1 - bs \in P$ and therefore both $s(1 - bs)$ and bs^2 are in R. This puts $s \in R$, which then implies that $bR + P = R$. Thus P is invertible. Moreover, as b was an arbitrary element of $PP^{-1}\backslash P = R\backslash P$, P is also a maximal ideal of R. □

Lemma 2.4.5. (Lucas [59, Theorem 2]) *Let R be a TP-domain. If P is a nonzero prime of R such that $P^{-1} \supsetneq (P : P)$, then P is an invertible maximal ideal of R.*

Proof. Suppose $P^{-1} \supsetneq (P : P)$ and let $b \in PP^{-1}\backslash P$. Also, let $I := b^2R + P$. Then $(R : I) = (P : P)$ (Lemma 2.4.3). Since R is a TP-domain, $b \in II^{-1}$ and therefore there are elements $s \in I^{-1} = (P : P)$ and $p \in P$ such that $b = b^2s + p$. Hence P is an invertible maximal ideal by Lemma 2.4.4. □

The fourth (and final) lemma appears in both [48] and [59].

Lemma 2.4.6. (Lucas [59, Lemma 14] and Houston–Kabbaj–Lucas–Mimouni [48, Theorem 3.4]) *Let P and P' be nonzero primes of a domain R. If $(R : P) = (P : P)$ and $(R : P') = (P' : P')$, then $(R : P \bigcap P') = (P \bigcap P' : P \bigcap P')$.*

Proof. Let $t \in (R : P \bigcap P')$ and set $J := P \bigcap P'$. Then $tJ \subseteq R$, $tP \subseteq (R : P') = (P' : P')$ and $tP' \subseteq (R : P) = (P : P)$. Thus J contains both tJP and tJP'. Hence $t^2JP \subseteq R$ and $t^2JP' \subseteq R$, and therefore $t^2J \subseteq (P : P) \bigcap (P' : P')$. It follows that $(tJ)^2 \subseteq J$ and we conclude with $tJ \subseteq J$ since J is a radical ideal of R (and $tJ \subseteq R$). □

Theorem 2.4.7. *Let R be a TP-domain. If R has at least one nonzero noninvertible ideal, then the nonmaximal primes are linearly ordered, and there is a (unique) prime ideal that contains all noninvertible ideals.*

Proof. Suppose I is a nonzero noninvertible ideal of R. Since R is a TP-domain, $Q := II^{-1}$ is a prime ideal of R. Also, $Q^{-1} = (Q : Q)$ (no matter whether R is a TP-domain or not). Therefore the set $\mathscr{X} := \{P \in \text{Spec}(R) \mid P^{-1} = (P : P)\}$ is nonempty. Clearly, $\mathscr{X}$ contains each noninvertible maximal ideal (if any). Moreover, by Lemma 2.4.5, each nonzero nonmaximal prime ideal of R is in $\mathscr{X}$.

To see that the primes in $\mathscr{X}$ are linearly ordered, let $P, P' \in \mathscr{X}$. Then $(R : P) = (P : P)$ and $(R : P') = (P' : P')$. By Lemma 2.4.6, $(R : P \bigcap P') = (P \bigcap P' : P \bigcap P')$. As R is a TP-domain, $P \bigcap P'$ is a prime ideal and hence P and P' are comparable. It follows that the primes of $\mathscr{X}$ are linearly ordered.

Let N be the union of the primes in $\mathscr{X}$. Then N is a prime that contains every nonmaximal prime of R. Let $t \in (R : N)$. Then $t \in (R : P) = (P : P)$ for each $P \in \mathscr{X}$. It follows that $tN \subseteq N$, which puts N in $\mathscr{X}$. From above, the prime $Q = II^{-1}$ is in $\mathscr{X}$. Hence we have $I \subseteq Q \subseteq N$. Note that if N is not maximal, then each maximal ideal of R is invertible. On the other hand, if N is maximal, then all other maximal ideals (if any) are invertible. □

In case of Prüfer ##-domains, a converse of Theorem 2.4.7 holds:

Proposition 2.4.8. (Fontana–Huckaba–Papick [23, Theorem 4.2]) *Let R be a ##-domain. If R is a Prüfer domain in which the noninvertible prime ideals are linearly ordered, then R is a TP-domain.*

In the Prüfer domains setting, the concept of RTP-domain provides a unified framework for studying ##-domains and TP-domains. To demonstrate this, we first recall a related trace property introduced by Lucas in 1996 [59, page 1095]. An integral domain R has the *trace property for primary ideals* (for short, R is a *TPP-domain*) if, for each nonzero primary ideal Q, either QQ^{-1} is prime or $QQ^{-1} = R$. By [59, Corollary 8], if R is a TPP-domain and Q is a primary ideal, then either $QQ^{-1} = \sqrt{Q}$ or $QQ^{-1} = R$ and $\sqrt{Q}$ is maximal.

Proposition 2.4.9. (Lucas [59, Theorem 4 and Corollary 8]) *Let R be an integral domain. Then:*

(1) If R is an RTP-domain then R is a TPP-domain.
(2) R is a TPP-domain if and only if, for each primary ideal Q of R, either $QQ^{-1} = P$, where $P := \sqrt{Q}$, or $QQ^{-1} = R$ and P is a maximal ideal of R.

A proof of Proposition 2.4.9 can also be found in [24, Theorem 4.2.17].

The converse of Proposition 2.4.9(1) is not known, except in certain special cases; e.g., Noetherian domains, or more generally, Mori domains [59, Theorem 12]. What is more interesting in our setting is that the converse holds for Prüfer domains.

Theorem 2.4.10. (Lucas [59, Theorem 23]) *Let R be a Prüfer domain. Then the following statements are equivalent.*

(i) R is an RTP-domain.
(ii) R is a TPP-domain.
(iii) Each nonzero branched prime of R is the radical of a finitely generated ideal (or, equivalently, by Proposition 2.3.10, each nonzero branched prime of R is sharp).

A proof of Theorem 2.4.10 can also be found in [24, Theorem 4.2.27]. More recently, S. El Baghdadi and S. Gabelli have shown that a Prüfer domain is an RTP-domain if and only if each nonzero principal ideal (equivalently, finitely generated ideal) has only finitely many minimal primes [18, Corollary 1.9].

The next result describes the relation between RTP-domains and ##- domains in the Prüfer domain setting.

Theorem 2.4.11. (Lucas [59, Corollaries 24, 25, and 26]) *Let R be a Prüfer domain.*

(1) If R is an RTP-domain, then every overring of R is a RTP-domain.
(2) If R is ##-domain, then R is a RTP-domain.
(3) If R is an RTP-domain and every maximal ideal of R is branched, then R is a #-domain.
(4) If R is an RTP-domain and every nonzero prime ideal of R is branched, then R is a ##-domain.

A proof of Theorem 2.4.11 also appears in [24, Theorem 4.2.28].

The next result is fundamental; it establishes the relationship among the classes of h-local, ##-, and RTP-domains in the Prüfer domain setting.

Theorem 2.4.12. (Olberding [67]) *Let R be a Prüfer domain. Then the following statements are equivalent.*

(i) R is h-local.
(ii) R is an RTP-domain and each nonzero prime ideal is contained in a unique maximal ideal.
(iii) R is a (##)-domain and each nonzero prime ideal is contained in a unique maximal ideal.
(iv) Each nonzero prime ideal of R is both sharp and contained in a unique maximal ideal.

The equivalences (i)⇔(ii)⇔(iii) are proved in [67, Proposition 3.4]). To see that (iii) and (iv) are equivalent, recall that by Theorem 2.2.4 (and the discussion which follows it), a Prüfer domain satisfies (##) if and only if each nonzero prime is sharp.

In the Prüfer domain case, using some of the preceding results, we can describe explicitly what is required to ensure that an RTP-domain is a TP-domain.

Corollary 2.4.13. *Let R be a Prüfer domain. The following statements are equivalent.*

(i) R is a TP-domain.
(ii) R is an RTP-domain and the noninvertible prime ideals of R are linearly ordered.
(iii) Each nonzero branched prime of R is the radical of a finitely generated ideal (or, equivalently, each nonzero branched prime of R is sharp), and the noninvertible prime ideals of R are linearly ordered.

Proof. Clearly, (ii)⇔(iii) by Theorem 2.4.10.

To see that (i) implies (ii) simply note that a TP-domain is a RTP-domain. That the primes are linearly ordered follows from Theorem 2.4.7.

To complete the proof assume (ii) holds. It is well known that if P is a minimal prime ideal of a nonzero ideal I of an integral domain and if I^{-1} is a ring, then P^{-1} is also a ring [48, Proposition 2.1]. Therefore, if J is a radical ideal of R such that J^{-1} is a ring, then it is easy to see that J must be a prime ideal. Hence R is a TP-domain. □

We also want to recall that the notions of RTP-domain and ##-domain often coincide in the Prüfer domain case:

Theorem 2.4.14. *Let R be a Prüfer domain.*

(1) (Heinzer–Papick [46, Theorem 2.7]) *Assume that R satisfies the ascending chain condition (for short, acc) on prime ideals (e.g., R is locally finite dimensional). The following statements are equivalent.*

   (i) *R is an RTP-domain.*
   (ii) Spec(R) *is a Noetherian space* (i.e., every nonempty set of closed subspaces of Spec(R) has a minimal element with respect to inclusion [10, Ch. II, §4, N. 2, Définition 3]).
   (iii) *R is a ##-domain.*

(2) (Gilmer–Heinzer [35, Theorem 4]) *The following statements are equivalent.*

   (i) *R is a ##-domain and R satisfies the acc on prime ideals.*
   (ii) *Every prime ideal of R is the radical of a finitely generated ideal (or, equivalently, by Proposition 2.3.10, every nonzero prime ideal of R sharp and branched).*

A proof of Theorem 2.4.14 also appears in [24, Theorems 4.2.33 and 4.2.34].

Putting together some of the equivalent statements of Proposition 2.1.8 and Theorem 2.4.12 (providing several characterizations for h-local domains), we have that *for a Prüfer domain R, R is an RTP-domain and each nonzero prime ideal is contained in a unique maximal ideal if and only if* $(R : I)R_M = (R_M : IR_M)$ *for each nonzero ideal I and each maximal ideal M.*

Our next result deals with what happens when the ideals I in the previous statement are restricted to primary ideals (or those that are locally primary).

Theorem 2.4.15. *The following statements are equivalent for a Prüfer domain R.*

(i) *R is an RTP-domain.*
(ii) $Q^{-1}R_M = (QR_M)^{-1}$ *for each nonzero primary ideal Q and each maximal ideal M of R.*
(iii) $I^{-1}R_M = (IR_M)^{-1}$ *for each nonzero locally primary ideal I of R and each maximal ideal M.*
(iv) *For each maximal ideal M of R,* $I^{-1}R_M = (IR_M)^{-1}$ *whenever* IR_M *is a nonzero primary ideal of* R_M.

Proof. Since R is Prüfer, it has the radical trace property if and only if each nonzero branched prime is the radical of a finitely generated ideal; i.e., if and only if each nonzero branched prime is sharp (Theorem 2.4.10). On the other hand, each nonzero branched prime is sharp if and only if $Q^{-1}R_M = (QR_M)^{-1}$ for each primary ideal Q of R, by Theorem 2.3.17. Hence (i) and (ii) are equivalent.

For the rest, it is clear that (iv) implies (iii) and (iii) implies (ii). Assume (ii) and let I and M be ideals with M maximal and IR_M primary. Simply let $Q := IR_M \bigcap R$. Then Q is a primary ideal that contains I and $QR_M = IR_M$. Hence we have $I^{-1}R_M \subseteq (IR_M)^{-1} = (QR_M)^{-1} = Q^{-1}R_M \subseteq I^{-1}R_M$. Thus (ii) implies (iv), completing the proof. □

According to Theorem 2.4.12, if R is a Prüfer domain R in which each nonzero prime is contained in a unique maximal ideal, then R is h-local if and only if it has the radical trace property. Hence we have the following corollary to Theorem 2.4.15.

Corollary 2.4.16. *The following statements are equivalent for a Prüfer domain R.*

(i) R is h-local.
(ii) Each nonzero prime is contained in a unique maximal ideal and $Q^{-1}R_M = (QR_M)^{-1}$ *for each primary ideal Q and maximal ideal M.*
(iii) Each nonzero prime is contained in a unique maximal ideal and for each nonzero ideal I and each maximal ideal M, $I^{-1}R_M = (IR_M)^{-1}$ *whenever* IR_M *is primary.*

In [19, page 3], we introduced a slightly weaker type of trace property for primary ideals: An integral domain R has the *weak trace property for primary ideals* (or, R is a *wTPP-domain*) if, for each nonzero primary ideal Q with nonmaximal radical, $QQ^{-1} = \sqrt{Q}$. With a slight modification of the proof that a Prüfer domain is a TPP-domain if and only if each branched prime is the radical of a finitely generated ideal (Theorem 2.4.10), we have the following.

Theorem 2.4.17. *Let R be a Prüfer domain. Then R is a wTPP-domain if and only if each branched nonmaximal prime is the radical of a finitely generated ideal.*

Proof. Since R is Prüfer, $PP^{-1} = P$ for each nonzero nonmaximal prime P (Theorem 2.3.2(2)). Also, by Theorem 2.3.12(1), if P is branched and nonmaximal, then P is the radical of a finitely generated ideal if and only if $QQ^{-1} = P$ for each proper P-primary ideal Q. □

A simple consequence of Theorem 2.4.17 is that if R is a Prüfer domain with the weak trace property for primary ideals, then each overring of R also has the weak trace property for primary ideals.

By [19, Proposition 1.7], a Prüfer domain with weak factorization is a wTPP-domain. It turns out that such a domain has a slightly stronger trace property. We say that an integral domain R *has the almost radical trace property* (or, is an *aRTP-domain*) if for each nonzero noninvertible ideal I of R, $II^{-1}R_M$ is a radical ideal, whenever M is either a steady maximal ideal or an unsteady maximal ideal that is not minimal over II^{-1}. In Theorem 4.2.12 below, we will show that a Prüfer domain R has weak factorization if and only if it is an aRTP-domain such that each nonzero prime is contained in a unique maximal ideal and each nonzero nonunit is contained in at most finitely many noninvertible maximal ideals.

Theorem 2.4.18. *The following are equivalent for a Prüfer domain R.*

(i) R is an aRTP-domain.
(ii) For each nonzero primary ideal Q, $QQ^{-1} \supseteq \sqrt{Q}$ except when $\sqrt{Q}$ is an unsteady maximal ideal.
(iii) Each nonzero branched nonmaximal prime ideal is the radical of a finitely generated ideal as is each steady branched maximal ideal.
(iv) Each nonzero branched nonmaximal prime ideal is sharp as is each steady branched maximal ideal.

Proof. We start with (i) implies (ii). Assume that R is an aRTP-domain and let Q be a nonzero primary ideal with radical P. If P is not maximal, then $QQ^{-1} \subseteq P$ (Lemma 2.3.15(3)). Hence in this case, $QQ^{-1}R_M$ is a radical ideal which must then be PR_M for each maximal ideal M. It follows that $QQ^{-1} = P$. If P is a steady maximal ideal, then by aRTP, we have $QQ^{-1}R_P \supseteq PR_P$ which means that either $QQ^{-1} = P$ or Q is invertible.

To see that (ii) implies (iii), first note that by Lemma 2.3.15, $QQ^{-1} \subseteq \sqrt{Q}$ whenever Q is a nonzero primary ideal such that $\sqrt{Q}$ is not maximal. Thus a Prüfer domain whose nonzero primary ideals satisfy statement (2) has the weak trace property for primary ideals. Hence, by Theorem 2.4.17, each nonzero branched nonmaximal prime is the radical of a finitely generated ideal. If Q is primary with $\sqrt{Q}$ a steady branched maximal ideal, then having $QQ^{-1} \supseteq \sqrt{Q}$ implies that $\sqrt{Q}$ is the radical of a finitely generated ideal by Theorem 2.3.12(1). Thus (ii) implies (iii).

The equivalence of (iii) and (iv) follows immediately from Proposition 2.3.10.

To complete the proof, we show that (iii) implies (i). Assume that each branched nonmaximal prime ideal is the radical of a finitely generated ideal, as is each steady branched maximal ideal. Let I be a nonzero noninvertible ideal with $J := II^{-1}$ and let M be a maximal ideal of R. If M does not contain J, then $JR_M = R_M$ is trivially a radical ideal of R_M. Assume that M contains J, and if M is minimal over J, then it is steady. Let $P \subseteq M$ be a prime ideal that is minimal over J. If P is unbranched, then $PR_M = JR_M$ since P has no proper primary ideals. On the other hand, if P is branched, then $PR_M = JR_M$ by Theorem 2.3.17. In either case, JR_M is a radical ideal of R_M. □

The "opposite" of a maximal ideal M being steady is for it to be unsteady, meaning MR_M is principal but M is not invertible. For Prüfer domains we have the following characterization of locally principal maximal ideals.

Lemma 2.4.19. *Let M be a maximal ideal of a Prüfer domain R. If M is unsteady, then M is not sharp. Equivalently, if M is sharp and locally principal, then M is invertible.*

Proof. We may assume that $MR_M = aR_M$ for some $a \in M$. If M is also sharp, then it is the radical of a finitely generated ideal I. By checking locally, we have $M = aR + I$, which is invertible. □

Since the property of being sharp is preserved in overrings (provided the prime survives), a sharp prime of a Prüfer domain cannot become unsteady in an overring. Thus, as with the weak trace property for primary ideals, if R is a Prüfer domain that is an aRTP-domain, then each overring is an aRTP-domain.

2.5 Sharp Primes and Intersections

Recall that each ideal in an overring of a Prüfer domain R is extended from an ideal of R [34, Theorem 26.1]. In particular, if T is an overring of R (with R Prüfer), then each maximal ideal of T is extended from a prime ideal of R. For a nonzero ideal I of the Prüfer domain R, we can say even more about the maximal ideals of the overring $\Gamma_R(I)$ $(:= \Gamma(I) := \bigcap\{R_M \mid M \in \text{Max}(R, I)\})$.

Lemma 2.5.1. *Let R be a Prüfer domain and let I be a nonzero ideal of R.*

(1) $\text{Max}(\Gamma(I)) = \{M\Gamma(I) \mid M \in \text{Max}(R, I)\}$.
(2) If $\sqrt{I} = P$ is a prime ideal and P' is the largest prime that is common to all maximal ideals that contain I, then $\Gamma(I) = \Gamma(P) = \Gamma(P')$ and $IR_{P'}$ is an ideal of $\Gamma(P)$. Moreover, $IR_{P'} = B\Gamma(P)$ where $B := IR_{P'} \bigcap R$.

Proof. Since R is Prüfer, each prime ideal of its overring $\Gamma(I)$ is extended from a unique prime ideal of R, and each prime of R that survives in $\Gamma(I)$ extends to a prime ideal of $\Gamma(I)$. From this and from the definition of $\Gamma(I)$, it is clear that each maximal ideal in the set $\text{Max}(R, I)$ extends to a maximal ideal of $\Gamma(I)$. Let Q be a prime ideal of R that is comaximal with I. Then there are elements $q \in Q$ and $a \in I$ such that $q + a = 1$. Clearly, no maximal ideal in $\text{Max}(R, I)$ contains q. Thus $1/q \in R_N$ for each $N \in \text{Max}(R, I)$, and hence $1/q \in \Gamma(I)$. Therefore $Q\Gamma(I) = \Gamma(I)$ and we have that each maximal ideal of $\Gamma(I)$ is extended from a maximal ideal in the set $\text{Max}(R, I)$, proving (1).

For (2), we further assume that $\sqrt{I} = P$ is a prime ideal and that P' is the largest prime that is common to all maximal ideals that contain P. In this case, a maximal ideal contains I if and only if it also contains P' (and P). Thus $\Gamma(I) = \Gamma(P) = \Gamma(P')$ and, by the first statement, the maximal ideals of $\Gamma(P)$ are precisely the ideals obtained by extending each maximal ideal of the set $\text{Max}(R, I) = \text{Max}(R, P) = \text{Max}(R, P')$. Since R_M contains $P'R_M = P'R_{P'}$ for each $M \in \text{Max}(R, P')$, $P'R_{P'}$ is an ideal of $\Gamma(P)$. Thus $IR_{P'}$ is also an ideal of $\Gamma(P)$.

From the definition of B, it follows that $BR_{P'} = IR_{P'}$, and for each $M \in \text{Max}(R, P)$, we have $BR_M = (BR_{P'} \bigcap R)R_M = BR_{P'}R_M \bigcap R_M = BR_{P'} \bigcap R_M = BR_{P'}$. Therefore $B\Gamma(P) = BR_{P'} = IR_{P'}$. □

Statements (2) and (3) in the next result generalize [19, Theorem 1.10]. The characterization of J^v in the case that P is sharp (also in (2)) is more or less from the proof of Proposition 4.1.3 (see [19, Theorem 1.10]), but here is given in a different form.

Theorem 2.5.2. *Let I be a finitely generated nonzero ideal of a Prüfer domain R and let P be a prime minimal over I and $J := IR_M \bigcap R$ where M is a maximal ideal that contains P.*

(1) If P is not sharp, then $J \subsetneq J^v = P^v = (R : \Theta(P))$.
(2) If P is sharp, then there is a finitely generated ideal B with $I \subseteq B$, $\sqrt{B} = P$, $J = BR_M \bigcap R$ and $J^v = J(P' : P') = B(P' : P')$ where P' is the largest prime which is contained in all the maximal ideals that contain P. Moreover, $J^{-1} = B^{-1}P'$ and $J \subsetneq J^v$ whenever more than one maximal ideal contains P.
(3) If P is sharp and M is the only maximal ideal that contains P, then J is divisorial and invertible.
(4) If $P = M$, then J is divisorial if and only if P is sharp.
(5) If M is not the only maximal ideal that contains P, then J is not divisorial.

Proof. Since P is minimal over I and I is finitely generated, $Q := I^2R_P \bigcap R$ is a P-primary ideal such that $Q = QR_M \bigcap R \subseteq IR_M \bigcap R = J$ (note that $IR_P \subseteq PR_P = PR_M$, whence $I^2R_P \subseteq IR_M$). Hence $\text{Max}(R, Q) = \text{Max}(R, J) = \text{Max}(R, P)$, and $\Theta(Q) = \Theta(J) = \Theta(P)$.

For (1), if P is not sharp, then $\Theta(P) = \Theta(Q) \supseteq Q^{-1} \supseteq J^{-1} \supseteq P^{-1} \supseteq R_P \bigcap \Theta(P) = \Theta(P)$, the last inclusion following from Theorem 2.3.2. It follows that $Q^v = J^v = P^v = (R : \Theta(P))$. Clearly, $J^v = R$ if P is maximal. If P is not maximal, then $J \subsetneq P \subseteq J^v$ (otherwise, $IR_M = JR_M = PR_M$, with I finitely generated and $P \subsetneq M$, which is impossible). Thus, in either case, J is not divisorial.

The statement in (4) follows from (1) and (3), and the statement in (5) follows from (1) and (2). Thus, for the remainder of the proof, we assume that P is sharp, whence by Proposition 2.3.10 there is a finitely generated ideal B with $\sqrt{B} = P$. Let P' be the largest prime which is contained in all the maximal ideals that contain P. Then P' is sharp by Lemma 2.3.9 since $\text{Max}(R, B) = \text{Max}(R, P) = \text{Max}(R, P')$.

Since P is minimal over I, $\sqrt{I}R_M = PR_M = \sqrt{B}R_M$ and thus there is an integer m such that $IR_M \supseteq B^mR_M$. Set $A := B^m$. Then $J = (I + A)R_M \bigcap R$ and $P = \sqrt{I + A}$. Without loss of generality, we may assume $B = I + A$ and thus $J = BR_M \bigcap R$.

The first statement in (2) is easier to prove when M is the only maximal ideal that contains P. In this case, $P' = M$, and M is the only maximal ideal that contains B. Thus $J = B$ (check locally) which implies J is invertible and therefore divisorial, establishing (3). Moreover, $(M : M) = R$ since M is a maximal ideal and R is Prüfer. Hence we trivially have $J^v = J = J(P' : P') = B(P' : P')$ in this case.

To complete the proof, we must show that $J^v = J(P' : P') = B(P' : P')$, $J \subsetneq J^v$, and $J^{-1} = B^{-1}P'$ when P is sharp and M is not the only maximal ideal that contains P. In this case, P' is properly contained in M. Thus $JR_M = BR_M \subsetneq JR_{P'} = BR_{P'}$ as B is finitely generated and R_M is a valuation domain. Since P' is sharp, $P'^{-1} = (P' : P') = R_{P'} \bigcap \Theta(P') \subsetneq \Theta(P')$ (Theorem 2.3.2(2)). Moreover, P' is a divisorial ideal of R and a maximal ideal of $(P' : P')$ (Corollary 2.3.21(2)). Also, $B(P' : P') = BP'^{-1}$ is a (proper) divisorial ideal of R with $(B(P' : P'))^{-1} = B^{-1}P'$ since B is invertible and P' is divisorial.

Let N be a maximal ideal. If N does not contain P, then R_N contains $(P' : P')$ and $BR_N = R_N = JR_N$. On the other hand, if N contains P, then it properly contains P' and blows up in $(P' : P')$ (as P' is a maximal ideal of $(P' : P')$). Thus $(P' : P')R_N$ contains $(P' : P')$ and properly contains R_N. Therefore there is a prime ideal $P_0 \subsetneq N$ such that $(P' : P')R_N = R_{P_0}$. But we also have $P_0 \subseteq P'$ and hence $P_0 = P'$. It follows that $B(P' : P')R_N = BR_{P'} = JR_{P'} = J(P' : P')R_N \supseteq JR_N$ when N contains P. Thus $B(P' : P') = J(P' : P')$ (since we have equality locally) is a divisorial ideal of R that contains J and the containment is proper since $JR_M = BR_M \subsetneq BR_{P'} \subseteq JR_{P'} = J(P' : P')R_M$. We also have $J^v \subseteq B(P' : P') = BP'^{-1}$. Since B is invertible, $J^vB^{-1} \subseteq P'^{-1}$ which implies $J^{-1}B \supseteq P'^v = P'$.

To finish the proof, it suffices to show that $J^{-1}B \subseteq P'$. For this, let $N(\neq M)$ be a maximal ideal that contains P, and let Q be the largest prime common to N and M. Then $JR_N = (BR_M \bigcap R)R_N = BR_Q$, and so for each $a \in N \backslash Q$, $a^{-1}B \subseteq BR_Q = JR_N$. Hence $a^{-1}BJ^{-1} \subseteq JJ^{-1}R_N \subseteq R_N$ and $BJ^{-1} \subseteq aR_N$. Therefore, using the fact that $B \subseteq J$, we have $BJ^{-1} \subseteq (\bigcap\{aR_N \mid a \in N \setminus Q\}) \bigcap R = QR_N \bigcap R = Q$. Now, let $L \subseteq M$ be a prime minimal over $J^{-1}B$. It suffices to show that $L \subseteq P'$. If not, then there is an N as above with $L \nsubseteq N$. For the corresponding Q we have $J^{-1}B \subseteq Q \subseteq M$. Since R is Prüfer, it must be the case that $L \subseteq Q \subseteq N$, a contradiction. Hence $L \subseteq P'$, as desired. □

Lemma 2.5.3. *Let R be a Prüfer domain, and let I be a nonzero ideal of R. If M is a steady maximal ideal that contains I where $I \subseteq I^vM \subsetneq I^v$, then M is idempotent and I^vR_M is principal.*

Proof. Suppose that M is a steady maximal ideal with $I \subseteq I^vM \subsetneq I^v$. Then it is also the case that $I^vMR_M \subsetneq I^vR_M$. Since M is steady, it is either idempotent or invertible (Lemma 2.1.10(1)). If M is invertible, then I^vM is a divisorial ideal that contains I and is properly contained in I^v. This is impossible, so it must be that M is idempotent. Let $y \in I^vR_M \setminus I^vMR_M$. Then $IR_M \subseteq I^vMR_M \subsetneq yR_M \subseteq I^vR_M$, and, since MR_M is not divisorial, taking v's in R_M yields $I^vR_M = yR_M$. □

Our next result "individualizes" two of the properties that characterize h-local Prüfer domains. Our ultimate goal is to give a characterization of Prüfer domains with weak factorization that is similar to Olberding's characterizations of h-local Prüfer domains (Proposition 2.1.8(2) and Theorem 2.4.12).

Theorem 2.5.4. *Let R be a Prüfer domain, and let N be a maximal ideal of R.*

(1) $I^{-1}R_N = (IR_N)^{-1}$ for each nonzero ideal I if and only if each nonzero prime contained in N is sharp and contained in no other maximal ideal.

(2) *If NR_N is principal and $I^{-1}R_N = (IR_N)^{-1}$ for some nonzero ideal I, then $IR_N = I^vR_N = (IR_N)^v$.*

(3) *If N is idempotent and $I^{-1}R_N = (IR_N)^{-1}$ for each nonzero ideal I, then $I^vR_N = (IR_N)^v$ for each nonzero ideal I, and for those I where $IR_N \neq I^vR_N$, $IR_N = yNR_N$ and $I^vR_N = yR_N$ for some $y \in I^v$.*

Proof. For (1), first assume each nonzero prime contained in N is sharp and contained in no other maximal ideal. Then $\Theta(N) = \Theta(P)$, and both N and P blow up in $\Theta(N)$ for each nonzero prime P contained in N. Thus $\Theta(N)R_N$ is the quotient field of R.

Let I be a nonzero ideal of R. Then $I^{-1} = (R_N : I)\bigcap(\Theta(N) : I)$ (Lemma 2.3.13). Thus $I^{-1}R_N = ((R_N : I)\bigcap(\Theta(N) : I))R_N = (R_N : I)R_N \bigcap(\Theta(N) : I)R_N = (IR_N)^{-1}$.

For the converse, we first show that if $I^{-1}R_N = (IR_N)^{-1}$ for each nonzero ideal I, then each nonzero branched prime contained in N is sharp. To this end, let $P \subseteq N$ be a branched prime and let Q be a proper P-primary ideal. Then $QQ^{-1}R_N = (QR_N)(QR_N)^{-1} \supsetneq PR_N$ since each valuation domain has the trace property (Proposition 2.4.1). It follows that $QQ^{-1} \supseteq P$. Moreover, if P is nonmaximal, then we have $QQ^{-1} = P$ by Lemma 2.3.15. Hence P is sharp by Theorem 2.3.12.

Next we show that no other maximal ideal contains P. Since P is both branched and sharp, there is a finitely generated ideal $B \subsetneq P$ such that $\sqrt{B} = P$ by Proposition 2.3.10. Let $J := BR_N \bigcap R$. By way of contradiction, assume that N is not the only maximal ideal that contains P. Then $J^{-1} = B^{-1}P'$ where P' is the largest prime common to all maximal ideals that contain P (Theorem 2.5.2(2)). Hence $J^{-1}R_N = B^{-1}P'R_N$. But $JR_N = BR_N$, so $(JR_N)^{-1} = (BR_N)^{-1} = B^{-1}R_N$ since B is finitely generated. Thus $JJ^{-1}R_N = P'R_N \subsetneq R_N = (JR_N)(JR_N)^{-1}$ and we have, upon canceling $JR_N = BR_N$, that $J^{-1}R_N$ is properly contained in $(JR_N)^{-1}$, a contradiction. Hence N is the only maximal ideal that contains P. This, in turn, implies that each nonzero prime contained in N is sharp and N is the only maximal ideal that contains it (Lemma 2.3.9).

For (2), assume NR_N is principal and $I^{-1}R_N = (IR_N)^{-1}$ for some nonzero ideal I. Since each nonzero ideal of R_N is divisorial (in R_N) (Lemma 2.1.1), we then have $IR_N \subseteq I^vR_N \subseteq (I^{-1}R_N)^{-1} = (IR_N)^v = IR_N$, and so $IR_N = I^vR_N = (IR_N)^v$.

Finally, for (3), assume that $I^{-1}R_N = (IR_N)^{-1}$ for each nonzero ideal I. This extends easily to all fractional ideals of R. Hence, for each nonzero ideal I, we have $I^vR_N = (I^{-1})^{-1}R_N = (I^{-1}R_N)^{-1} = (IR_N)^v$.

For N idempotent, the existence of the element $y \in I^v$ such that $IR_N = yNR_N$ and $I^vR_N = yR_N$ when $IR_N \neq I^vR_N(= (IR_N)^v)$ follows from the fact that all nondivisorial ideals of R_N have the form zNR_N for some $z \in R$ (Lemma 2.1.1). □

Recall that for a nonzero ideal I of an integral domain R, $\Phi(I) = \bigcap\{R_P \mid P \in \mathrm{Min}(R, I)\}$, where $\mathrm{Min}(R, I)$ is the set of minimal primes of I (in R), and $\Theta(I) = \bigcap\{R_N \mid N \in \mathrm{Max}(R) \setminus \mathrm{Max}(R, I)\}$.

Lemma 2.5.5. *Let R be a Prüfer domain and let I be a nonzero radical ideal of R such that I^{-1} is a ring.*

(1) If P is both minimal over I and antesharp, then PI^{-1} is a maximal ideal of I^{-1}.
(2) If P' is a sharp prime that contains I and $P'I^{-1} \neq I^{-1}$, then P' is minimal over I.

Note that the statement in (2) is not a consequence of that in (1). Simply having I contained in a sharp prime P' with P' not minimal over I is not enough to guarantee that the minimal prime $P \subsetneq P'$ of I is antesharp.

Proof. (1) Assume that P is minimal over I and antesharp. Since R is Prüfer and I^{-1} is a ring, $I^{-1} = \Theta(I) \bigcap \Phi(I) \subseteq R_P$ (Theorem 2.3.2(1)). Thus PI^{-1} is a prime ideal of I^{-1}. If P is maximal in R, then PI^{-1} is certainly maximal in I^{-1}. Otherwise, since P is antesharp, each prime P'' that properly contains P also contains an invertible ideal B that (properly) contains P (Proposition 2.3.20). Since $I \subseteq P \subsetneq B$, $B^{-1} \subseteq I^{-1}$. As B is invertible, we have $I^{-1} \supseteq P''I^{-1} \supseteq BI^{-1} = I^{-1}$. (To see that $BI^{-1} = I^{-1}$, write $1 = \sum b_i u_i$ with $b_i \in B$ and $u_i \in B^{-1} \subseteq I^{-1}$, whence $1 \in BI^{-1}$.) Therefore $P''I^{-1} = I^{-1}$ for each prime P'' that properly contains P, and so PI^{-1} is maximal ideal of I^{-1}.

For (2), assume that P' is a sharp prime that contains I. We will show that if P' is not minimal over I, then $P'I^{-1} = I^{-1}$. So we further assume P' is not minimal over I and let $P \subsetneq P'$ be minimal over I. If P' is unbranched, there is an infinite chain of sharp branched primes between P and P' (Theorem 2.3.19). Thus, whether P' is unbranched or branched, there is a sharp branched prime P'' with $I \subseteq P \subsetneq P'' \subseteq P'$. Inside P'' is a finitely generated ideal A such that $\sqrt{A} = P''$ (Proposition 2.3.10). Obviously, if N is a maximal ideal that does not contain I, then N does not contain A. Thus $\Theta(A) \subseteq \Theta(I)$. Since $A^{-1} \subseteq \Omega(A) \subseteq \Theta(A)$ (Lemma 2.3.1), then $A^{-1} \subseteq \Theta(I)$. Moreover, no prime minimal over I can contain A, so $A^{-1} \subseteq \Phi(I)$ as well. It follows that $A^{-1}I^{-1} = \Phi(I) \bigcap \Theta(I) = I^{-1}$ and therefore $I^{-1} = AI^{-1} \subseteq P'I^{-1}$, as desired. □

Recall from Remark 2.3.6 that Gilmer and Heinzer showed that given a fixed nonzero prime P and a nonempty set of primes $\{P_\alpha\}$ in a Prüfer domain R, R_P does not contain $T := \bigcap_\alpha R_{P_\alpha}$ if and only if there is a finitely generated ideal $A \subseteq P$ that is contained in no P_α. Also, if the primes in $\{P_\alpha\}$ are pairwise incomparable and, in our terminology, each P_α is sharp, then each $P_\alpha T$ is a sharp maximal ideal of T and no other maximal ideals of T (if any) are sharp. We now consider the related problem of which (minimal) primes are essential in an intersection of incomparable primes.

Theorem 2.5.6. *Let I be a radical ideal of a Prüfer domain R that is not prime, and let $\{P_\alpha\}$ be a set of minimal primes of I such that $I = \bigcap_\alpha P_\alpha$. If Q is a minimal prime of I that is sharp, then $Q = P_\beta$ is in the set $\{P_\alpha\}$, and $\bigcap_{\alpha \neq \beta} P_\alpha$ properly contains I, as does $P \bigcap (\bigcap_{\alpha \neq \beta} P_\alpha)$ for each prime P that properly contains Q.*

Proof. Let Q be a minimal prime of I that is sharp. Then there is a finitely generated ideal $A \subseteq Q$ such that the only maximal ideals that contain A are those that contain Q (Proposition 2.3.5(1)). If M is such a maximal ideal, then $IR_M = QR_M \supseteq AR_M$. Since A is finitely generated, M does not contain the ideal $J := (I :_R A)$. On the other hand, if $P_\gamma \in \{P_\alpha\}\backslash\{Q\}$, then P_γ must contain J since it contains I and cannot contain A. Thus $I \subsetneq J \subseteq \bigcap_{P_\alpha \neq Q} P_\alpha$, and therefore $Q = P_\beta$ for some β.

Next, assume P is a prime that properly contains Q. Since $J \nsubseteq Q$, the product PJ is not contained in Q but is contained in P and also in the intersection $\bigcap_{\alpha \neq \beta} P_\alpha$. Thus we also have $I \subsetneq P \bigcap (\bigcap_{\alpha \neq \beta} P_\alpha)$. □

A useful corollary to Theorem 2.5.6 is the following.

Corollary 2.5.7. *Let R be a Prüfer domain with nonzero Jacobson radical J, and let P be a sharp prime that is minimal over J. Then the following statements are equivalent.*

(i) P is a maximal ideal.
(ii) P is contained in a unique maximal ideal.
(iii) P is contained in only finitely many maximal ideals.

Proof. Let $\{P_\alpha\}$ be the set $\text{Min}(R, J) \setminus P$, and let $B = \bigcap_\alpha P_\alpha$. We start by proving (ii) implies (i). Let M be the maximal ideal that contains P. If $P \neq M$, then $M \bigcap B \supsetneq J$ by Theorem 2.5.6. But since $\text{Spec}(R)$ is treed, no P_α is contained in M. Thus $B \subseteq I := \bigcap\{N \mid N \in \text{Max}(R, P_\alpha) \text{ for some } \alpha\}$, and we have $J \subsetneq M \bigcap B \subseteq M \bigcap I = J$, a contradiction. Hence $P = M$ is maximal.

To see that (iii) implies (i), we revisit the proof of Theorem 2.5.6. Assume P is not maximal but is contained in only finitely many maximal ideals $M_1, M_2, \ldots, M_n$ (with $n > 1$). Since P is sharp, there is a finitely generated ideal $A \subseteq P$ with $\text{Max}(R, A) = \text{Max}(R, P) = \{M_1, M_2, \ldots, M_n\}$ (Proposition 2.3.5). No M_i contains the ideal $C := (J :_R A)$ since A is finitely generated and $AR_{M_i} \subsetneq PR_{M_i} = JR_{M_i}$. Hence C is not contained in P and thus neither is $CM_1M_2\cdots M_n$. On the other hand, $C \subseteq B$ since no P_α contains A. Thus $CM_1M_2\cdots M_n$ is contained in each maximal ideal, a contradiction. Therefore P is a maximal ideal of R. □

Theorem 2.5.8. *Let R be a Prüfer domain and let I be a nonzero radical ideal. If each minimal prime of I is a maximal ideal of R and $I = \bigcap\{N \mid N \in \mathscr{W}\}$ for some subset $\mathscr{W}$ of* $\text{Max}(R, I)$*, then $\Gamma(I) = \bigcap\{R_N \mid N \in \mathscr{W}\}$.*

Proof. Assume that each minimal prime of I is a maximal ideal of R and let $\mathscr{W}$ be a subset of $\text{Max}(R, I)$ such that $I = \bigcap\{N \mid N \in \mathscr{W}\}$. There is nothing to prove if $\mathscr{W} = \text{Max}(R, I)$, so suppose $M \notin \mathscr{W}$ is a maximal ideal that contains I. To simplify notation, set $T := \bigcap\{R_N \mid N \in \mathscr{W}\}$. By way of contradiction, assume that R_M does not contain T. Then there is an element $q \in T$ that is not in R_M. It follows that M must contain the invertible ideal $C := (R : (1, q))$. On the other hand, no $N \in \mathscr{W}$ contains C.

Since each maximal ideal in $\mathrm{Max}(R, I)$ is minimal over I and I is a radical ideal, $IR_M = MR_M \supseteq CR_M$. Since C is finitely generated, the ideal $E := (I :_R C)$ is not contained in M. On the other hand, $EC \subseteq I \subseteq N$ for each $N \in \mathscr{W}$. This leads to the contradictory statement that $E \subseteq \bigcap\{N \mid N \in \mathscr{W}\} = I \subseteq M$ since, as observed above, no $N \in \mathscr{W}$ contains C. Therefore we must have $R_M \supseteq T$ for each maximal ideal $M \in \mathrm{Max}(R, I)$, and it follows that $\Gamma(I) = T$. □

Let $\mathscr{S} := \{P_\alpha \mid \alpha \in \mathscr{A}\}$ be a nonempty set of incomparable nonzero prime ideals of a Prüfer domain R. We say that a prime $P_\beta \in \mathscr{S}$ is *relatively sharp in* $\mathscr{S}$ if P_β contains a finitely generated ideal that is contained in no other prime in $\mathscr{S}$. If $\mathscr{S}$ is a finite set, then each prime in $\mathscr{S}$ is relatively sharp (in $\mathscr{S}$), vacuously so if $\mathscr{S}$ has only one member. Since R is Prüfer, P_β is relatively sharp in $\mathscr{S}$ (assuming $\mathscr{S}$ contains more than one prime) if and only if R_{P_β} does not contain the intersection $\bigcap\{R_{P_\alpha} \mid \alpha \in \mathscr{A}, \alpha \neq \beta\}$ (Remark 2.3.6). If each $P_\alpha \in \mathscr{S}$ is relatively sharp in $\mathscr{S}$, then $\mathscr{S}$ is said to be a *relatively sharp set*.

As with Theorem 2.5.8, the next two results deal with intersections of maximal ideals. In the first, R is a Prüfer domain and the maximal ideals are not sharp in R but are all relatively sharp to each other. In the second, the maximal ideals are sharp in R but there is no Prüfer assumption.

Theorem 2.5.9. *Let R be a Prüfer domain, let I be a nonzero radical ideal of R such that each minimal prime of I is a maximal ideal of R, and let $\mathscr{W} := \{M_\alpha \mid \alpha \in \mathscr{A}\}$ be a subset of* $\mathrm{Max}(R, I)$. *If $I = \bigcap\{M_\alpha \mid \alpha \in \mathscr{A}\}$ and each M_α is unsteady but relatively sharp in $\mathscr{W}$, then $I^{-1} = R$.*

Proof. If $I = \bigcap\{M_\alpha \mid \alpha \in \mathscr{A}\}$ with each M_α unsteady (and so, locally principal) but relatively sharp in $\mathscr{W}$, then each M_α contains a finitely generated ideal J_α of R that is contained in no other maximal ideal in $\mathscr{W}$. Since M_α is locally principal, we may further assume that $J_\alpha R_{M_\alpha} = M_\alpha R_{M_\alpha}$.

By Theorem 2.5.8, $\Gamma(I) = \bigcap\{R_{M_\alpha} \mid \alpha \in \mathscr{A}\}$. To simplify the notation, we set $T := \Gamma(I)$ and $T_\beta := \bigcap\{R_{M_\alpha} \mid \alpha \in \mathscr{A}, \alpha \neq \beta\}$ for each $\beta \in \mathscr{A}$. Clearly, $T_\beta = \bigcap\{T_{M_\alpha T} \mid \alpha \in \mathscr{A}, \alpha \neq \beta\}$, $T = T_\beta \bigcap T_{M_\beta T}$ and $J_\beta T_{M_\beta T} = M_\beta T_{M_\beta T}$ for each $\beta \in \mathscr{A}$.

Since $J_\beta^{-1} \subseteq R_{M_\alpha}$ for each $\alpha \neq \beta$, $J_\beta^{-1} \subseteq T_\beta$. It follows that both J_β and M_β blow up in T_β. By Lemma 2.3.13, $(T : M_\beta T) = (T_{M_\beta T} : M_\beta T_{M_\beta T}) \bigcap (T_\beta : M_\beta T_\beta) = (T_{M_\beta T} : M_\beta T_{M_\beta T}) \bigcap T_\beta = (T_{M_\beta T} : J_\beta T_{M_\beta T}) \bigcap T_\beta = (T : J_\beta T)$. Thus $M_\beta T = J_\beta T$ is an invertible maximal ideal of T, which makes it sharp as well.

Consider the ideal $I_\beta := \bigcap\{M_\alpha \mid \alpha \in \mathscr{A}, \alpha \neq \beta\}$. Since $IR_{M_\beta} = M_\beta R_{M_\beta} = J_\beta R_{M_\beta}$ and J_β is finitely generated, M_β does not contain the ideal $(I :_R J_\beta)$, but all other M_α's do. Clearly, $J_\beta I_\beta \subseteq I$. Thus $I_\beta = (I :_R J_\beta)$ and since it is not contained in M_β, then $M_\beta + I_\beta = R$. Therefore there are elements $c_\beta \in M_\beta$ and $d_\beta \in I_\beta$ such that $c_\beta + d_\beta = 1$. Also $I = M_\beta I_\beta = M_\beta \bigcap I_\beta$. Without loss of generality we may assume $c_\beta \in J_\beta$. Thus $J_\beta + I_\beta = R$ as well.

For each β, consider the ring $\Gamma(J_\beta)$. By Lemma 2.5.1, having $J_\beta + I_\beta = R$ implies $I_\beta \Gamma(J_\beta) = \Gamma(J_\beta)$. Thus $M_\beta \Gamma(J_\beta) = I\Gamma(J_\beta)$ since $I = M_\beta I_\beta$.

On the other hand, since M_β is not sharp in R (Lemma 2.4.19), it must fail to be sharp in $\Gamma(J_\beta)$ and therefore its inverse in $\Gamma(J_\beta)$ is trivial (Remark 2.1.2(1)). Hence $(\Gamma(J_\beta) : I\Gamma(J_\beta)) = \Gamma(J_\beta)$. With this we have $(R : I) \subseteq \Gamma(J_\beta) \subseteq R_{M_\beta}$ and therefore $(R : I) \subseteq \bigcap\{R_{M_\alpha} \mid \alpha \in \mathscr{A}\} = T$. That $(R : I) = R$ now follows from the facts that $T \bigcap \Theta(I) = \Gamma(I) \bigcap \Theta(I) = R$ and $(R : I) \subseteq \Theta(I)$ (Lemma 2.3.1). □

Theorem 2.5.10. *Let R be an integral domain and let $\mathscr{M} = \{M_\alpha\}$ be a set of sharp maximal ideals of R. If $R = \bigcap_\alpha R_{M_\alpha}$, then the only sharp maximal ideals of R are those in the set $\mathscr{M}$, and $\bigcap_\alpha M_\alpha$ is the Jacobson radical of R.*

Proof. There is nothing to prove if $\mathscr{M} = \operatorname{Max}(R)$, so assume there is a maximal ideal M that is not one of the M_α's. Clearly, if $R = \bigcap_\alpha R_{M_\alpha}$, then M is not sharp. Moreover, for $t \in \bigcap_\alpha M_\alpha$, if t is not in M, then there are elements $p \in R$ and $q \in M$ such that $pt + q = 1$. It follows that q is a unit in each R_{M_α}. But having $R = \bigcap_\alpha R_{M_\alpha}$ implies q is a unit of R, a contradiction. Thus $t \in M$, and $\bigcap_\alpha M_\alpha$ is the Jacobson radical of R. □

Another useful theorem is the following. In particular, it allows us to give an alternate proof of Theorem 2.1.6.

Theorem 2.5.11. *Let R be a Prüfer domain with nonzero Jacobson radical J.*

(1) If each maximal ideal of R is invertible and minimal over J, then R has only finitely many maximal ideals.
(2) If each maximal ideal of R is invertible and each nonzero prime is both sharp and contained in a unique maximal ideal, then R has only finitely many maximal ideals.

Proof. Let $\operatorname{Max}(R) = \{M_\beta\}$ and assume each M_β is invertible.

For (1), assume each M_β is minimal over J. Also, for each β, let $C_\beta := \bigcap_{\alpha \neq \beta} M_\alpha$. Since each M_β is invertible, each is sharp and therefore M_β is comaximal with C_β. It follows that $R = \sum_\beta C_\beta$. Hence there are maximal ideals $M_{\beta_1}, M_{\beta_2}, \ldots, M_{\beta_n}$ such that $R = C_{\beta_1} + C_{\beta_2} + \cdots + C_{\beta_n}$. As $C_\beta \subseteq M_\alpha$ for all $\alpha \neq \beta$, $M_{\beta_1}, M_{\beta_2}, \ldots, M_{\beta_n}$ are the only maximal ideals of R.

For (2), assume each nonzero prime is both sharp and contained in a unique maximal ideal. Since R is a Prüfer domain, each M_β contains a unique prime P_β that is minimal over J, and, since each prime is contained in a unique maximal ideal, the P_β are distinct.

For each β, let $J_\beta := \bigcap_{\alpha \neq \beta} P_\alpha$ and $I_\beta := \bigcap_{\alpha \neq \beta} M_\alpha$. Since P_β is sharp, J_β properly contains J by Theorem 2.5.6 and P_β does not contain J_β. Since $J \subseteq M_\beta \bigcap J_\beta \subseteq M_\beta \bigcap I_\beta = J$, Theorem 2.5.6 further implies $P_\beta = M_\beta$. Hence each maximal is minimal over J. By (1), R has only finitely many maximal ideals. □

Simply having a nonzero Jacobson radical with each maximal ideal invertible and each nonzero prime ideal sharp is not enough to imply only finitely many maximal ideals. For example, it is well known that the domain $\mathbb{Z} + X\mathbb{Q}[[X]]$ is a two-dimensional Prüfer domain such that each maximal ideal is invertible and the Jacobson radical is the (unique nonzero) nonmaximal prime $X\mathbb{Q}[[X]]$. Clearly, there are infinitely many maximal ideals and each nonzero prime is sharp (since the only such nonmaximal prime is contained in every maximal ideal), but no maximal ideal is minimal over the Jacobson radical.

A rather unusual characterization of Prüfer domains involves a certain nice "factoring" property of contents of polynomials. For an indeterminate X over R, the *content* of a polynomial $h \in R[X]$ is the ideal of R generated by the coefficients of R. We shall use $\mathbf{c}(h)$ to denote this ideal. It turns out that R is a Prüfer domain if and only if $\mathbf{c}(fg) = \mathbf{c}(f)\mathbf{c}(g)$ for all nonzero polynomials $f, g \in R[X]$ [34, Theorem 28.6].

Here is an alternate proof of Heinzer's characterization of integrally closed divisorial domains (Theorem 2.1.6).

Theorem 2.5.12. (Heinzer [44, Theorem 5.1]) *Let R be an integrally closed integral domain. Then each nonzero ideal of R is divisorial if and only if R is an h-local Prüfer domain such that each maximal ideal is invertible.*

Proof. Assume R is an h-local Prüfer domain such that each maximal ideal is invertible. Let M be a maximal ideal and let I be a nonzero ideal contained in M. Since M is invertible, MR_M is principal and IR_M is divisorial (Lemma 2.1.1(5)). Also, each nonzero prime is contained in a unique maximal ideal, and from finite character, each such prime is sharp. Thus $I^{-1}R_M = (IR_M)^{-1}$ by Theorem 2.5.4. Hence $IR_M = (IR_M)^v = I^vR_M$. It follows that $I = I^v$.

For the converse, assume R is an integrally closed domain such that each nonzero ideal is divisorial. Since R is integrally closed, $\mathbf{c}(fg)^v = (\mathbf{c}(f)\mathbf{c}(g))^v$ for each pair of nonzero polynomials $f, g \in R[X]$ [34, Proposition 34.8]. Hence we have $\mathbf{c}(fg) = \mathbf{c}(fg)^v = (\mathbf{c}(f)\mathbf{c}(g))^v = \mathbf{c}(f)\mathbf{c}(g)$. Thus R is a Prüfer domain [34, Theorem 28.6]. That each maximal ideal is invertible follows from the fact a maximal ideal of a Prüfer domain is divisorial if and only if it is invertible.

Let P be a branched prime, and let M be a maximal ideal that contains P. Then P is minimal over a finitely generated ideal I. Since $IR_M \bigcap R$ is divisorial, Theorem 2.5.2 ensures that P is sharp and that M is the only maximal ideal that contains P. As each (nonzero) unbranched prime contains a branched prime, each nonzero prime is sharp and contained in a unique maximal ideal.

Next, let $r \in R$ be a nonzero nonunit. Then the only maximal ideals of $\Gamma(rR)$ are those that are extended from maximal ideals of R that contain r (Lemma 2.5.1(1)). Thus the Jacobson radical of $\Gamma(rR)$ is nonzero, each maximal ideal of $\Gamma(rR)$ is invertible, and each nonzero prime of $\Gamma(rR)$ is both sharp and contained in a unique maximal ideal. By Theorem 2.5.11, $\Gamma(rR)$ has only finitely many maximal ideals. Therefore R has finite character and hence is h-local. □

We end this section with three lemmas which will be used later.

Lemma 2.5.13. *The following are equivalent for a nonzero prime P of a Prüfer domain R.*

(i) *P is invertible as an ideal of $(P : P)$.*
(ii) *P is sharp and PR_P is principal.*
(iii) *There is an invertible prime of $(P : P)$ that contains P.*

Proof. If P is a maximal ideal, then $(P : P) = R$. In this case, P invertible is equivalent to it being sharp and locally principal. For the remainder of the proof we assume P is not maximal.

If P is invertible as an ideal of $(P : P)$, then it is a maximal ideal of $(P : P)$ and there is a finitely generated ideal $I \subseteq P$ such that $P(P : P) = I(P : P)$. As each maximal ideal of R that does not contain P extends to a maximal ideal of $(P : P)$, no such maximal ideal contains I. Hence P is sharp and PR_P is principal. Thus (i) implies (ii).

If P is sharp and PR_P is principal, then P is branched and a maximal ideal of $(P : P)$ (Theorem 2.3.11 and Corollary 2.3.21). Also, $P = \sqrt{J}$ for some finitely generated ideal J, which we may further assume has the property that $JR_P = PR_P$. The other maximal ideals of $(P : P)$ are extended from maximal ideals of R that do not contain P [24, Theorem 3.1.2]. Hence checking locally in $(P : P)$ yields $P(P : P) = J(P : P)$. Thus (ii) implies both (i) and (iii).

To complete the proof, we show that (iii) implies (i). We prove the contrapositive. Suppose P is not invertible as an ideal of $(P : P)$ and let $Q \supsetneq P$ be a prime of R. Since R is a Prüfer domain, $PQ = P$ and $(R : P) = (P : P)$. It follows that $((P : P) : Q) = ((R : P) : Q) = (R : PQ) = (R : P) = (P : P)$. Hence each prime of $(P : P)$ that properly contains P has a trivial inverse in $(P : P)$ and so is not invertible. □

Recall that an ideal I is said to be *SV-stable* (*stable in the sense of Sally-Vasconcelos*) if I is invertible as ideal of $(I : I)$. Also recall that if $(R : I) = (I : I)$ for some ideal I, then $(R : P) = (P : P)$ for each minimal prime P of I [48, Proposition 2.1(2)].

Lemma 2.5.14. *Let R be a domain, and let I be an ideal of R. If each minimal prime of I is an invertible maximal ideal of R, then not only is I invertible but so is each ideal that contains I.*

Proof. It suffices to prove the following form of the contrapositive: if some ideal that contains I is not invertible but each minimal prime of I is maximal, then some minimal prime of I is not invertible. Let J be a noninvertible ideal that contains I and let $B := J(R : J)$. Then $I \subseteq B$ and $(R : B) = (B : B)$, and so by [48, Proposition 2.1], each minimal prime P of B is such that $(R : P) = (P : P)$. Hence no such prime is invertible. It follows that some minimal prime of I is not invertible. □

Lemma 2.5.15. *Let I be a radical ideal of a Prüfer domain R. If $(R : I) = (I : I)$ and each minimal prime of I is SV-stable, then I has only finitely many minimal primes and I is SV-stable.*

Proof. Assume $(R : I) = (I : I)$ and each minimal prime of I is SV-stable. To simplify notation, we let $T := (I : I)$. Also, let $P \in \mathrm{Min}(R, I)$. Since $(R : I) = (I : I) = T$, $T \supseteq (R : P) = (P : P)$ [48, Proposition 2.1(2)]. As P is invertible as an ideal of $(P : P)$, it is maximal as an ideal of $(P : P)$. Moreover, $R_P \supseteq T \supseteq (P : P)$, whence PT is an invertible maximal ideal of T. Thus $\mathrm{Max}(T, I) = \{P(I : I) \mid P \in \mathrm{Min}(R, I)\}$ with each maximal ideal in $\mathrm{Max}(T, I)$ invertible. It follows from Lemma 2.5.14 that I is invertible as an ideal of T, equivalently, I is SV-stable. We also have $\Gamma_T(I) = \Phi_R(I)$ with each maximal ideal of $\Gamma_T(I)$ invertible. By Theorem 2.5.11, $\Gamma_T(I)$ has only finitely many maximal ideals. Hence I has only finitely many minimal primes. □

Chapter 3
Factoring Ideals in Almost Dedekind Domains and Generalized Dedekind Domains

Abstract We start with an overview of the rings for which every proper ideal is a product of radical ideals, rings introduced by Vaughan and Yeagy under the name of SP-rings. The integral domains with this property are called here domains with radical factorization. We give several characterizations of this type of integral domains by revisiting, completing and generalizing the work by Vaughan–Yeagy (Canad. J. Math. 30:1313–1318, 1978) and Olberding (Arithmetical properties of commutative rings and monoids, Chapman & Hall/CRC, Boca Raton, 2005). In Sect. 3.2, we study almost Dedekind domains having the property that each nonzero finitely generated ideal can be factored as a finite product of powers of ideals of a factoring family (definition given below). In the subsequent section, we provide a review of the Prüfer domains in which the divisorial ideals can be factored as a product of an invertible ideal and pairwise comaximal prime ideals, after papers by Fontana–Popescu (J. Algebra 173:44–66, 1995), Gabelli (Commutative Ring Theory, Marcel Dekker, New York, 1997) and Gabelli–Popescu (J. Pure Appl. Algebra 135:237–251, 1999). The final section is devoted to the presentation of various general constructions due to Loper–Lucas (Comm. Algebra 31:45–59, 2003) for building examples of almost Dedekind (non Dedekind) domains of various kinds (e.g., almost Dedekind domains which do not have radical factorization or which have a factoring family for finitely generated ideals or which have arbitrary sharp or dull degrees (definitions given below)).

3.1 Factoring with Radical Ideals

In a 1978 paper, Vaughan and Yeagy [75] studied the rings for which every proper ideal is a product of radical ideals (also called *semi-prime ideals*or *SP-ideals*). The rings with this property were called *SP-rings*. To emphasize the factorization aspect, we will say that an integral domain with this property has *radical factorization*. In addition, we say that an ideal I has a *radical factorization* if there are finitely many radical ideals $J_1, J_2, \ldots, J_n$ such that $I = J_1 J_2 \cdots J_n$.

M. Fontana et al., *Factoring Ideals in Integral Domains*, Lecture Notes of the Unione Matematica Italiana 14, DOI 10.1007/978-3-642-31712-5_3,

By one of the main results of Vaughan–Yeagy, an SP-domain is an almost Dedekind domain [75, Theorem 2.4]. Using a construction given by Heinzer and J. Ohm [45] of a non-Noetherian almost Dedekind domain, Vaughan and Yeagy also provide a non-Noetherian example of an SP-domain.

In 1976, Butts and Yeagy [14] introduced the notion of a maximal ideal being *critical* as meaning that each finite subset of this particular maximal ideal is contained in the square of some maximal ideal. In 1979, Yeagy used this concept to give another class of non trivial examples of SP-domains as an application of the following.

Theorem 3.1.1. (Yeagy [77, Theorem 3.2]) *Let R be an almost Dedekind domain that is a union of a tower of Dedekind domains. Then R is an SP-domain if and only if R has no critical maximal ideals.*

Clearly, no maximal ideal of a Dedekind domain is critical since a finitely generated maximal ideal is obviously not critical. Also, an idempotent maximal ideal is obviously critical.

The classical example of a non-Dedekind almost Dedekind domain constructed by N. Nakano in 1953 [65], using the ring of integers in the (non finite) number field obtained by adjoining to $\mathbb{Q}$ the p-th roots of unity, for all primes p, turns out to be an SP-domain since it has no critical maximal ideals.

An example of an almost Dedekind domain that is a union of a tower of Dedekind domains having a critical maximal ideal was given by Butts and Yeagy in 1976 [14]. In particular, this example shows that SP-domains are a proper subclass of the class of almost Dedekind domains.

In 2005, Olberding [69] completed and generalized Vaughan–Yeagy results, proving several characterizations of SP-domains. We list some of these in the following.

Theorem 3.1.2. (Olberding [69, Theorem 2.1]) *Let R be an integral domain, but not a field. The following statements are equivalent.*

(i) R is a SP-domain; i.e., R has radical factorization.
(ii) R is an almost Dedekind domain having no critical maximal ideals.
(iii) R is an almost Dedekind domain and, for every proper finitely generated ideal J of R, $\sqrt{J}$ is finitely generated.
(iv) R is a one-dimensional Prüfer domain, and every proper principal ideal of R is the product of radical ideals.
(v) Every proper nonzero ideal I of R can be uniquely represented as a product $I = Q_1Q_2\cdots Q_n$ of radical ideals Q_k, where $Q_1 \subseteq Q_2 \subseteq \cdots \subseteq Q_n$.
(vi) R is a one-dimensional Prüfer domain having no critical maximal ideals.

Note that the equivalences (i)⇔(ii)⇔(iii)⇔(v)⇔(vi) are given explicitly in [69, Theorem 2.1((i)⇔(ii)⇔(iii)⇔(vii)) and Corollary 2.2]. The equivalence (ii)⇔(vi) is a straightforward consequence of the fact that in a one-dimensional Prüfer domain having no critical maximal ideals, $M \neq M^2$ for all $M \in \operatorname{Max}(R)$, and this implies that MR_M is a principal ideal in the one-dimensional valuation domain R_M, for all $M \in \operatorname{Max}(R)$.

Vaughan and Yeagy also introduced a function to "measure" a nonzero ideal in an almost Dedekind domain. Olberding gives an alternate characterization of SP-domains (= domains with radical factorization) in terms of continuity properties of this function [69, Theorem 2.1].

We provide an alternate characterization of radical factorization below that more or less mimics what Olberding did, but without explicit mention of continuity properties.

Before presenting this result, we revisit a notion considered by Butts and Gilmer in 1964 [13].

An integral domain R is said to have *property* (α) if every primary ideal is a power of its radical [13]. It is clear that if R is a domain with property (α), then both R_P and R/P have property (α) for each prime P of R.

The next three lemmas are based on [13, Theorems 1, 2 and 3]. The first is just a local version of the second.

Lemma 3.1.3. *Let R be a local domain with property (α). If the maximal ideal M is minimal over an ideal of the form $tR + P$ for some $t \in M\backslash P$ and some nonmaximal prime P, then $\bigcap_n M^n$ is a prime ideal that contains P and is properly contained in M.*

Proof. If M is minimal over the ideal $I_1 := tR + P$, then I_1 is M-primary. The same conclusion holds for $I_k := t^k R + P$ for each $k \geq 1$. If $t^k = t^{k+1}s + q$ for some elements $s \in R$, $q \in P$, then $t^k(1 - ts) = q$ implies $q \in M\backslash P$, a contradiction. Hence $I_k \supsetneq I_{k+1}$ for each k. By property (α), for each k there is an integer $m_k \geq 1$ with $I_k = M^{m_k}$. Since $M^{m_k} = I_k \supsetneq I_{k+1} = M^{m_{k+1}}$, each power of M is distinct. Hence a consequence of property (α) is that $M^n = bR + M^m$ for each $b \in M^n\backslash M^{n+1}$ and all positive integers $m > n$.

Let $J := \bigcap_n M^n = \bigcap_k M^{m_k} \supseteq P$. Then $M \supsetneq J$ and for $x, y \in M\backslash J$, there are integers n and m such that $x \in M^n\backslash M^{n+1}$ and $y \in M^m\backslash M^{m+1}$. Thus $M^n = xR + M^{n+1}$ and $M^m = yR + M^{m+1}$. As $M^{n+m+1} \subsetneq M^{n+m} = xyR + xM^{m+1} + yM^{n+1} + M^{n+m+2}$, $xy \in M^{n+m}\backslash M^{n+m+1}$. Therefore J is a prime ideal of R that contains P and is properly contained in M. □

Lemma 3.1.4. *Let R be a domain with property (α), and let P be a prime of R. If Q is a prime minimal over an ideal $tR + P$ for some $t \notin P$, then $Q \supsetneq \bigcap_n Q^n \supseteq P$ with $\bigcap_n Q^n$ a prime ideal.*

Proof. Since R_Q has property (α) with QR_Q minimal over $tR_Q + PR_Q$, Lemma 3.1.3 implies that $\bigcap_n Q^n R_Q$ is a prime ideal of R_Q that contains PR_Q and is properly contained in QR_Q. For each integer $k \geq 1$, set $I_k := t^k R + P$. By first localizing at Q and then contracting to R, we obtain Q-primary ideals which by property (α) must be powers of Q. Let m_k be such that $I_k R_Q \bigcap R = Q^{m_k}$. Then by Lemma 3.1.3, $\bigcap_k I_k R_Q$ is a prime ideal of R_Q that contains PR_Q and is properly contained in QR_Q. Thus there is a prime $Q_0 \subsetneq Q$ such that $\bigcap_k I_k R_Q = Q_0 R_Q$ with $P \subseteq Q_0 \subseteq Q^n$ for each n. It follows that $Q_0 = \bigcap_n Q^n \subsetneq Q$. □

Lemma 3.1.5. *Let R be a domain with property (α) and let N be a nonzero prime ideal of R. Then $\bigcap_n N^n$ is a prime ideal of R that contains every prime that is properly contained in N. Moreover, if $N \neq N^2$, then NR_N is principal.*

Proof. Suppose P is a prime that is properly contained in N, and let $t \in N \backslash P$. Then there is a prime $Q \subseteq N$ that is minimal over $tR + P$. By Lemma 3.1.4, $\bigcap_n Q^n$ is a prime ideal that contains P and is properly contained in Q. Obviously, $\bigcap_n N^n$ contains $\bigcap_n Q^n$. It follows that $\bigcap_n N^n$ contains each prime that is properly contained in N. If $N = N^2$, we simply have $\bigcap_n N^n = N$. If $N \neq N^2$, then by property (α), $NR_N = bR_N + N^2R_N$ for each $b \in N \backslash N^2$. By the proof of Lemma 3.1.3, $\bigcap_n N^n$ is a prime ideal in this case as well.

Continuing with the assumption that $N \neq N^2$, set $Q_0 := \bigcap_n N^n$ and let $r \in N \backslash N^2$. Since Q_0 (which is properly contained in N^2) contains each prime that is properly contained in N and it is properly contained in N, NR_N is the radical of rR_N. Thus rR_N is NR_N-primary. By property (α), the only possibility is to have $NR_N = rR_N$. □

In the absence of property (α), it is possible to obtain the same conclusion as that in Lemma 3.1.5 under the assumption the "N" is the radical of a finitely generated ideal and $\{N^n\}$ is the complete set of N-primary ideals. We will find the next lemma useful in Sect. 5.3.

Lemma 3.1.6. *Let R be a local domain with maximal ideal M. If M is the radical of a finitely generated ideal, then M is principal if and only if $\{M^n \mid n \geq 1\}$ is the complete set of M-primary ideals. Moreover, if M is principal, then $\bigcap_n M^n$ is a nonmaximal prime ideal that contains each nonmaximal prime of R.*

Proof. It is well-known that if M is principal, then the only M-primary ideals are the powers of M (and each is distinct). For the converse, assume M is the radical of a finitely generated ideal I and that $\{M^n \mid n \geq 1\}$ is the complete set of M-primary ideals. As I is M-primary, there is an integer $n \geq 1$ such that $I = M^n$. Then $M^{2n} = I^2 \subsetneq I$, and therefore $M \supsetneq M^2 \supsetneq M^3 \supsetneq \cdots$. Moreover, we have $M^{n-1} = bR + M^n$ for each $b \in M^{n-1} \backslash M^n$. Hence M^{n-1}, and recursively, M^k is finitely generated for each $k \leq n$. In particular, M is finitely generated.

To see that M is principal, suppose, by way of contradiction, that M is minimally generated by $n > 1$ elements, say $M = (a_1, a_2, \ldots, a_n)$. Since $M \neq M^2$, we may assume that $a_n \notin M^2$. The ideal $(a_1^2, a_2^2, \ldots, a_{n-1}^2, a_n)$ is M-primary and must therefore be equal to M. This gives an equation $a_1 = r_1a_1^2 + r_2a_2^2 + \cdots + r_{n-1}a_{n-1}^2 + ra_n$. However, we then have $a_1(1 - r_1a_1) \in (a_2, \ldots, a_n)$, with $1 - r_1a_1$ a unit, contradicting that n is minimal. Hence M is indeed principal. Let $M = (a)$. Now suppose, again by way of contradiction, that for some nonmaximal prime Q and some $k \geq 1$, we have $Q \subseteq M^k$ but $Q \nsubseteq M^{k+1}$. Then $Q + M^{k+1}$ is M-primary, whence $Q + M^{k+1} = M^k$, and we have an equation $a^k = q + ta^{k+1}$, with $t \in R$, $q \in Q$. However, this yields $a^k(1 - ta) \in Q$, a contradiction. □

We now have enough to prove that a domain with radical factorization is an almost Dedekind domain.

Theorem 3.1.7. (Vaughan–Yeagy [75, Theorem 2.4]) *If R is a domain with radical factorization, then R is an almost Dedekind domain.*

Proof. Let Q be a proper P-primary ideal for some nonzero prime P and factor Q into radical ideals as $Q = J_1 J_2 \cdots J_n$. Since Q is P-primary, if some J_i is not contained in P, we have $Q \subseteq \prod_{j \neq i} J_j \subseteq Q$. On the other hand, if $J_j \subseteq P$, we must, in fact, have $J_j = P$ since each prime that contains Q also contains P. Hence $Q = P^k$ for some integer $k \geq 1$. Therefore R has property (α).

If P is a height one prime, then the only way a nonzero principal ideal of R_P can factor into radical ideals is as a power of PR_P. It follows that PR_P is principal. Thus we are done once we show R is one-dimensional.

By way of contradiction, assume $\dim(R) > 1$ and let $P \subsetneq N$ be a pair of nonzero prime ideals. We may further assume N is minimal over an ideal of the form $tR + P$ for some $t \in N \backslash P$ and that P is minimal over a (nonzero) principal ideal sR. Then $NR_N \neq N^2 R_N$ by Lemma 3.1.3, and thus by Lemma 3.1.5, NR_N is principal and $\bigcap_n N^n \subsetneq N$ contains each prime that is properly contained in N.

Factor the ideal sR_N as $sR_N = I_1 I_2 \cdots I_n R_N$ with each $I_i R_N$ a radical ideal of R_N. Each $I_i R_N$ is invertible and at least one, say $I_1 R_N$, is contained in PR_N. Then NR_N is not minimal over $I_1 R_N$ and by Lemma 2.4.3(2), we obtain a contradiction from having $R_N \subsetneq (R_N : NR_N) \subseteq (I_1 R_N : I_1 R_N) = R_N$. Therefore R is one-dimensional, as desired. □

Olberding's continuity characterization for radical factorization is based on the collection of functions $\{\gamma_a \mid a \in R\backslash\{0\}\}$ each of which maps $\operatorname{Max}(R)$ to $\mathbb{Z}$ as follows: $\gamma_a(M) := k$ where $aR_M = M^k R_M$ (equivalently $\gamma_a(M) := \mathrm{v}_M(a)$ where v_M is the valuation map corresponding to R_M). By [69, Theorem 2.1], an almost Dedekind domain R has radical factorization if and only if the function γ_a is both bounded and upper semi-continuous (in the Zariski topology on $\operatorname{Max}(R)$) for each nonzero $a \in R$. Note that γ_a is upper semi-continuous if and only if $\gamma_a^{-1}([n, \infty))$ is closed in $\operatorname{Max}(R)$. It turns out that simply having γ_a upper semi-continuous for each nonzero nonunit $a \in R$ is sufficient to imply that R has radical factorization (and thus that γ_a is bounded for each a). We take a slightly different approach in the next theorem.

For each nonzero ideal I and each maximal ideal M of an almost Dedekind domain R, we set $\rho_M(I) := h$ if $IR_M = M^h R_M$. Then we set $\rho(I) := \sup\{\rho_M(I) \mid M \in \operatorname{Max}(R)\}$. Also, for each positive integer k, let $\mathscr{Y}_k(I) := \{M \in \operatorname{Max}(R) \mid IR_M \subseteq M^k R_M\}$ and $I_k := \bigcap\{M \in \mathscr{Y}_k(I)\}$, with $I_k := R$ if $\mathscr{Y}_k(I)$ is empty. It is clear that $I_1 = \sqrt{I} \subseteq I_2 \subseteq \cdots$. We employ this notation in the statement of our next theorem. In the case $I = bR$ is a principal ideal, we simply use $\mathscr{Y}_k(b)$ in place of $\mathscr{Y}_k(I)$. The characterizations in (ii)–(v) are new.

Theorem 3.1.8. (Vaughan–Yeagy [75, Sect. 3] and Olberding [69, Theorem 2.1]) *The following statements are equivalent for an almost Dedekind domain R.*

(i) R has radical factorization.

(ii) For each nonzero ideal I, $\rho(I)$ is bounded and, for each positive integer $k \leq \rho(I)$, the only maximal ideals that contain I_k ($= \bigcap\{M \in \mathscr{Y}_k(I)\}$) are those in the corresponding set $\mathscr{Y}_k(I)$.

(iii) *For each nonzero finitely generated ideal J and each positive integer k, the only maximal ideals that contain $J_k := \bigcap\{N \in \mathcal{Y}_k(J)\}$ are those in the corresponding set $\mathcal{Y}_k(J)$.*

(iv) *For each nonzero principal ideal bR and each positive integer k, the only maximal ideals that contain $B_k := \bigcap\{N \in \mathcal{Y}_k(b)\}$ are those in the corresponding set $\mathcal{Y}_k(b)$.*

(v) *The function γ_a is upper semi-continuous for each nonzero nonunit $a \in R$.*

Proof. It is clear that (ii) implies (iii) and that (iii) implies (iv).

Let b be a fixed nonzero nonunit of R and for each $k \geq 1$, let $B_k := \bigcap\{N \in \mathcal{Y}_k(b)\}$ (equal to R if $\mathcal{Y}_k(b)$ is empty). It is clear that $\gamma_b^{-1}([k,\infty)) = \mathcal{Y}_k(b)$. If the only maximal ideals of R that contain B_k are those in the set $\mathcal{Y}_k(b)$, then $\mathcal{Y}_k(b)$ is closed. Conversely, if $\mathcal{Y}_k(b)$ is closed and M is a maximal ideal that contains B_k, then $M \in \mathcal{Y}_k(b)$. Hence (iv) and (v) are equivalent.

Next we show (iv) implies (ii).

Let I be a nonzero ideal of R. Then for each maximal ideal $M \in \text{Max}(R, I)$, there is a positive integer n such that $I \subseteq M^n$ and $I \not\subseteq M^{n+1}$. Thus there is an element $b \in I \backslash M^{n+1}$. As above, we let $B_k = \bigcap\{N \in \mathcal{Y}_k(b)\}$ and $I_k = \bigcap\{N \in \mathcal{Y}_k(I)\}$. We have $\mathcal{Y}_k(b) \supseteq \mathcal{Y}_k(I)$ for each k and so $B_k \subseteq I_k$. Since $b \notin M^{n+1}$, M does not contain B_{n+1} and thus does contain I_{n+1}. It follows that the only maximal ideals that contain I_{n+1} are those in the $\mathcal{Y}_{n+1}(I)$. Therefore for each k, the only maximal ideals that contain I_k are those in the set $\mathcal{Y}_k(I)$.

To see that $\rho(I)$ is bounded, note that the family of ideals $\{I_k\}_{k=1}^{\infty}$ forms an ascending chain. Hence $H := \bigcup_{k=1}^{\infty} I_k$ is an ideal of R. For a given maximal ideal N of R, there is an integer k such that $IR_N = N^k R_N$ which puts $N \in \mathcal{Y}_k(I) \backslash \mathcal{Y}_{k+1}(I)$. Thus $N \supseteq I_k$ but N does not contain I_{k+1} and therefore N does not contain H. It follows that $H = R$ and thus there is an integer m such that $I_m \subsetneq I_{m+1} = R$ which implies $\rho(I) = m$. Therefore (iv) implies (ii).

To see that (ii) implies (i), suppose that for each nonzero ideal I, $\rho(I)$ is bounded and for each positive integer $k \leq \rho(I)$, the only maximal ideals that contain I_k are those in the corresponding set $\mathcal{Y}_k(I)$. We will show that $I = I_1 I_2 \cdots I_m$ where $m = \rho(I)$.

Since $I_j \subseteq I_{j+1}$ for all j ($\leq m-1$), if a maximal ideal N contains I_{j+1}, it also contains I_j. As each I_j is a radical ideal, $I_j R_N = NR_N$ whenever N contains I_j and $I_j R_N = R_N$ when N does not contain I_j. By our assumptions on the ideals I_j, $I_j \subseteq N$ if and only if $IR_N \subseteq N^j R_N$. Hence if $\rho_N(I) = h$, then $N \supseteq I_j$ for all $j \leq h$ and $I_j R_N = R_N$ for all $j > h$. It follows that $IR_N = N^h R_N = I_1 I_2 \cdots I_h R_N$, with $I_{h+1} \cdots I_m R_N = R_N$ whenever $h < m$. Hence $IR_M = I_1 I_2 \cdots I_m R_M$ for each maximal ideal M, and thus $I = I_1 I_2 \cdots I_m$. Therefore R has radical factorization.

To complete the proof we show that (i) implies (iii). Note that if $\rho(I)$ is unbounded for some nonzero ideal I, then there is no hope of factoring I as a finite product of radical ideals (since if J is a radical ideal, we have $JR_M = MR_M \supsetneq M^2 R_M$ for each maximal ideal $M \supseteq J$). Hence if R has radical factorization, then $\rho(I) < \infty$ for each nonzero ideal I.

If B is a finitely generated ideal such that $\rho(B) = 1$, then $B = B_1$ is a radical ideal and $\mathcal{Y}_k(B)$ is empty for each $k > 1$. We proceed by induction on $\rho(J)$.

Assume (iii) holds for each finitely generated ideal B with $\rho(B) < m$ and let J be a finitely generated ideal with $\rho(J) = m$. Let $J = A_1 A_2 \cdots A_n$ be a factorization of J into radical ideals. Then each A_i is invertible. Obviously, each maximal ideal that contains J contains at least one A_i. Since R is a Prüfer domain, $A_1 \bigcap A_2 \bigcap \cdots \bigcap A_n$ is both an invertible ideal [34, Proposition 25.4] and a radical ideal. In fact, $A_1 \bigcap A_2 \bigcap \cdots \bigcap A_n = \sqrt{J} = J_1$ and the ideal $J(R : J_1)$ is invertible.

Let N be a maximal ideal that contains J. Then $JR_N = N^k R_N$ for some integer $1 \leq k \leq m$ and $NR_N = J_1 R_N$. Thus $J(R : J_1)R_N = N^{k-1}R_N$. It follows that $\mathcal{Y}_k(J) = \mathcal{Y}_{k-1}(J(R : J_1))$ and $\rho(J(R : J_1)) = m - 1$. By the induction hypothesis, the only maximal ideals that contain $\bigcap\{N \in \mathcal{Y}_j(J(R : J_1))\} = \bigcap\{P \in \mathcal{Y}_{j+1}(J)\} = J_{j+1}$ are those in $\mathcal{Y}_j(J(R : J_1)) = \mathcal{Y}_{j+1}(J)$. Therefore (i) implies (iii). □

Remark 3.1.9. It is still an open problem to find characterizations of SP-rings with zero divisors.

3.2 Factoring Families for Almost Dedekind Domains

As is well-known (and recalled above in Proposition 2.1.3), in a Dedekind domain, each nonzero proper ideal can be factored (uniquely) as a finite product of positive powers of maximal ideals. More generally, Gilmer proved the following lemma.

Lemma 3.2.1. [34, Proposition 37.5] *If I is a proper ideal of an almost Dedekind domain R that is contained in only finitely many maximal ideals $M_1, M_2, \ldots, M_n$, then $I = M_1^{e_1} M_2^{e_2} \cdots M_n^{e_n}$, for some positive integers $e_1, e_2, \ldots, e_n$.*

In 2003, A. Loper and Lucas [58] observed that a consequence of the previous lemma is that a finitely generated ideal in an almost Dedekind domain may have a factorization into prime ideals. More precisely, if $\mathrm{Max}^{\#}(R)$ denotes the set of sharp maximal ideals of R, they proved the following.

Proposition 3.2.2. (Loper–Lucas [58, Lemma 2.2]) *Let R be an almost Dedekind domain and let I be a finitely generated proper ideal of R. Then there are maximal ideals $M_1, M_2, \ldots, M_n$ of R such that $I = M_1^{e_1} M_2^{e_2} \cdots M_n^{e_n}$ for some positive integers $e_1, e_2, \ldots, e_n$ if and only if $I \nsubseteq N$, for all $N \in Max(R) \backslash Max^{\#}(R)$.*

Proof. First suppose there are maximal ideals $M_1, M_2, \ldots, M_n$ such that I factors as $I = M_1^{e_1} M_2^{e_2} \cdots M_n^{e_n}$, for some positive integers $e_1, e_2, \ldots, e_n$. Then, obviously, no other maximal ideals contain I. For each M_i, there is an element $b_i \in M_i \backslash \bigcup_{j \neq i} M_j$. The ideal $B_i := b_i R + I$ is a finitely generated ideal with radical M_i. Hence each M_i is sharp (and invertible).

For the converse, assume each maximal ideal that contains I is in $\mathrm{Max}^{\#}(R)$. Let M be a maximal ideal that contains I. Since MR_M is principal and M is sharp, M is finitely generated and therefore invertible. Also $IR_M = M^e R_M$ for some positive integer e. By Lemma 2.5.1, each maximal ideal of $\Gamma(I)$ is extended from a maximal ideal in the set $\text{Max}(R, I)$. It follows that each maximal ideal of $\Gamma(I)$ is both invertible and minimal over $I\Gamma(I)$. Hence by Theorem 2.5.11, $\Gamma(I)$ has only finitely many maximal ideals. Thus $\text{Max}(R, I)$ is finite and, for each $M_i \in \text{Max}(R, I)$, we have $IR_{M_i} = M_i^{e_i} R_{M_i}$ for some positive integer e_i. Checking locally shows that $I = M_1^{e_1} M_2^{e_2} \cdots M_n^{e_n}$. □

In the same paper [58], Loper and Lucas investigated the following problem: *Given an almost Dedekind domain R with $Max(R) = \{M_\alpha \mid \alpha \in \mathscr{A}\}$, when is it possible to find a family of finitely generated ideals $\{J_\alpha \mid \alpha \in \mathscr{A}\}$ of R such that $J_\alpha R_{M_\alpha} = M_\alpha R_{M_\alpha}$ for each α, and every finitely generated nonzero ideal of R can be factored as a finite product of powers of ideals from the family $\{J_\alpha \mid \alpha \in \mathscr{A}\}$?*

In order to state some of the main results obtained in [58] concerning this problem, we need some preliminary notions.

For an almost Dedekind domain R with $\text{Max}(R) = \{M_\alpha \mid \alpha \in \mathscr{A}\}$, we say that a set of finitely generated ideals $\mathscr{J} := \{J_\alpha \mid \alpha \in \mathscr{A}\}$ is a *factoring family* for R if $J_\alpha R_{M_\alpha} = M_\alpha R_{M_\alpha}$ for each α, and every finitely generated nonzero ideal of R can be factored as a finite product of powers of ideals from the family $\mathscr{J}$. A *factoring set* of an almost Dedekind domain is a factoring family such that no member appears more than once.

Given a one-dimensional Prüfer domain R with quotient field K, call a maximal ideal M *dull* if $M \in \text{Max}^\dagger(R) := \text{Max}(R) \backslash \text{Max}^\#(R)$. We can recursively define a family of overrings of R as follows:

$$R_1 := R, \quad R_n := \bigcap\{(R_{n-1})_N \mid N \in \text{Max}^\dagger(R_{n-1})\} \text{ for } n > 1,$$

where $R_n = K$, for $n \geq 2$, if $\text{Max}^\dagger(R_{n-1}) = \emptyset$.

We say that R has *sharp degree n* if $R_n \neq K$ but $R_{n+1} = K$ (and *dull degree n* if $R_{n-1} \subsetneq R_n = R_{n+1} \subsetneq K$, with $R_0 = \{0\}$). Note that a domain is a #–domain if and only if it has sharp degree 1. For an fractional ideal J of R, we say that J has *sharp degree n* if $JR_n \neq R_n$, but $JR_{n+1} = R_{n+1}$. Note that a proper (integral) ideal I of R has sharp degree 1 if and only if each maximal ideal containing I is sharp.

Theorem 3.2.3. (Loper–Lucas [58, Theorem 2.3 and Corollary 2.4]) *Let R be an almost Dedekind domain such that each maximal ideal has finite sharp degree. Then there exists a factoring set $\mathscr{J}$ such that each finitely generated fractional ideal of R factors uniquely over $\mathscr{J}$. In particular, every almost Dedekind domain of finite sharp degree admits a factoring set.*

The factoring set of this theorem can be simply constructed as follows. First note that the maximal ideals of R which generate sharp maximal ideals of R_n are exactly the maximal ideals of R of sharp degree n. For each maximal ideal M_α of R, we know that M_α has finite sharp degree, say n, with $n \geq 1$. Then pick J_α to be a finitely generated ideal of R such that $J_\alpha R_{M_\alpha} = M_\alpha R_{M_\alpha}$ and J_α is contained in no other maximal ideal of R_n. Finally, set $\mathscr{J} := \{J_\alpha \mid \alpha \in \mathscr{A}\}$.

Theorem 3.2.4. (Loper–Lucas [58, Theorem 2.5]) *Let R be a one-dimensional Prüfer domain. Then R is an almost Dedekind domain with at most one noninvertible maximal ideal if and only if there exists a nonzero element $d \in R$ such that, for each finitely generated nonzero ideal I of R, there is a finite set of maximal ideals $\{M_1, M_2, \ldots, M_m\}$, a finite set of integers $\{e_1, e_2, \ldots, e_m\}$, and a nonnegative integer n such that*

$$I = (d)^n M_1^{e_1} M_2^{e_2} \cdots M_m^{e_m}.$$

Moreover, if R is an almost Dedekind domain with exactly one noninvertible maximal ideal N, then the element $d \in R$ must be such that $dR_N = NR_N$, and the set $\mathscr{J} := Max^{\#}(R) \bigcup \{dR\}$ is a factoring set for R.

The paper [58] contains a general construction for obtaining almost Dedekind domains with various sharp degrees and, in particular, an almost Dedekind domain of sharp degree 2 satisfying the hypothesis of Theorem 3.2.4. It also includes a general construction for obtaining almost Dedekind domains with various dull degrees. In the construction, the resulting almost Dedekind domain R is the union of a countable chain of Dedekind domains $\{R_n\}$ such that for each maximal ideal M of R, $M_n R_M = MR_M$, where $M_n = M \bigcap R_n$ is a maximal (hence, finitely generated) ideal of R_n. It follows that R has no critical maximal ideals and therefore is an SP-domain (Theorems 3.1.1 or 3.1.2); equivalently, R has radical factorization. We consider this construction in more detail in Sect. 3.4

3.3 Factoring Divisorial Ideals in Generalized Dedekind Domains

A Prüfer domain R is said to be a *generalized Dedekind domain* if each localizing system is finitely generated (or, equivalently, $R_{\mathscr{F}} = R_{\mathscr{G}}$ for a pair of localizing systems $\mathscr{F}$ and $\mathscr{G}$ of R implies $\mathscr{F} = \mathscr{G}$) [73, Sect. 2], [72], and [24, Sect. 5.2]. In the local case, the generalized Dedekind domains can be characterized as follows.

Theorem 3.3.1. (Fontana–Popescu [26, Théorème 2.2]) *The following statements are equivalent for a domain R.*

(i) R is a local generalized Dedekind domain.
(ii) R is a valuation domain such that PR_P is principal for each nonzero prime P of R.
(iii) R is a discrete valuation domain (i.e., no branched prime ideal is idempotent, [34, page 192]) and each prime ideal of R is the radical of a principal ideal.
(iv) R is a valuation domain such that each ideal can be factored as a principal ideal times a prime ideal.

A valuation domain V with no nonzero idempotent prime ideals is said to be *strongly discrete*; equivalently, PV_P is principal for each nonzero prime ideal P. Similarly, a Prüfer domain R is *strongly discrete* if it has no nonzero idempotent prime ideals; equivalently, PR_P is principal for each nonzero prime P.

From the equivalence of (i) and (ii), a local generalized Dedekind domain is the same as a strongly discrete valuation domain. The analogous equivalence does not hold for generalized Dedekind domains. While a generalized Dedekind domain is a strongly discrete Prüfer domain, there are strongly discrete Prüfer domains that are not generalized Dedekind domains.

From a global point of view, we have the following characterizations of generalized Dedekind domains (see [24, Theorems 5.3.8 and 5.4.9]).

Theorem 3.3.2. (Popescu [73, Theorem 2.5]). *Let R be an integral domain. The following statements are equivalent.*

(i) R is a generalized Dedekind domain.
(ii) R is a Prüfer domain with no nonzero idempotent prime ideals and each nonzero prime of R is the radical of a finitely generated ideal.
(iii) R is strongly discrete Prüfer domain and $Spec(R)$ is a Noetherian space.

By Proposition 2.3.10 above, a nonzero branched prime of a Prüfer domain is sharp if and only if it is the radical of a finitely generated ideal. Also, by Theorem 2.4.10, each branched prime of a Prüfer domain R is the radical of a finitely generated ideal if and only R is an RTP-domain. Thus we have the following alternate characterization of generalized Dedekind domain. The equivalence of (i) and (ii) is due to Gabelli [28].

Corollary 3.3.3. (Gabelli [28, Theorem 5]) *The following statements are equivalent for a Prüfer domain R.*

(i) R is a generalized Dedekind domain.
(ii) R is an RTP-domain with no nonzero idempotent prime ideals.
(iii) R is a Prüfer domain such that each nonzero prime ideal is sharp, and no nonzero prime is idempotent.

Theorem 3.3.4. (Gabelli–Popescu [31, Theorem 3.3]) *Let R a generalized Dedekind domain and let I be a nonzero ideal of R. Then $I^v = HP_1P_2\cdots P_n$ for some invertible ideal H and prime ideals $P_1, P_2, \ldots, P_n$ of R.*

In the above mentioned paper, Gabelli and Popescu gave a more precise statement, providing a characterization of generalized Dedekind domains among Prüfer domains [31, Theorem 3.3]. For a domain R, we let $\mathrm{Div}(R)$ denote the set of divisorial (integral) ideals of R, $\mathbf{F}^v(R)$ denote the set of divisorial fractional ideals of R, $\mathrm{Inv}(R)$ the set of invertible ideals of R (including R itself) and $\mathbf{H}(R)$ the set of invertible fractional ideals of R. What Gabelli and Popescu showed is that a Prüfer domain R is a generalized Dedekind domain if and only if $\mathbf{F}^v(R) = \{BP_1P_2\cdots P_n \mid B \in \mathbf{H}(R)$ and $P_1, P_2, \ldots, P_n$ are pairwise comaximal primes$\}$. Their proof can be easily modified to show that one also has R a generalized Dedekind domain if and only if (R is Prüfer and) $\mathrm{Div}(R) = \{IQ_1Q_2\cdots Q_m \mid I \in \mathrm{Inv}(R)$ and $Q_1, Q_2, \ldots, Q_m$ are pairwise comaximal primes$\}$. We will use several of our earlier results to give a different proof for this equivalence.

The following lemma was observed in [25, page 495].

Lemma 3.3.5. *If I is a divisorial ideal of a domain R, then $(I : I) = (II^{-1} : II^{-1}) = (R : II^{-1})$.*

Proof. It is always the case that $(I : I) \subseteq (II^{-1} : II^{-1}) = (R : II^{-1})$. Let $t \in (R : II^{-1})$, $q \in I$ and $b \in (R : I)$. Since $(R : II^{-1}) \subseteq (R : I)$, both tq and $t(qb) = (tq)b$ are in R. Thus $tq \in I^v = I$. □

Theorem 3.3.6. *The following statements are equivalent for a Prüfer domain R.*

(i) R is a generalized Dedekind domain.
(ii) $\mathbf{F}^v(R) = \{HP_1P_2\cdots P_n \mid H \in \mathbf{H}(R)$, and $P_1, P_2, \ldots, P_n$ are pairwise comaximal prime ideals of $R\}$.
(iii) $Div(R) = \{IP_1P_2\cdots P_n \mid I \in Inv(R)$, and $P_1, P_2, \ldots, P_n$ are pairwise comaximal prime ideals of $R\}$.

We prove that (i) and (iii) are equivalent. For a proof of the equivalence of (i) and (ii), see [31].

Proof. Assume R is a generalized Dedekind domain. By Theorem 3.3.2, there are no nonzero idempotent primes and each nonzero prime is the radical of a finitely generated ideal and thus sharp. Hence each maximal ideal of R is invertible. For a nonzero prime P, PR_P is principal and P is sharp. Hence P is an invertible maximal ideal of $(P : P)$ by Lemma 2.5.13. Therefore by Lemma 2.5.15, if H is a radical ideal of R with $(R : H) = (H : H)$, then H has only finitely many minimal primes and H is invertible as an ideal of $(H : H)$.

If A is an invertible proper ideal of R and M is a maximal ideal that contains A, then AM^{-1} is an invertible ideal of R (equal to R if $A = M$) and we have $A = (AM^{-1})M$.

Next, let I be a noninvertible ideal of R and let $J := I(R : I)$. Then $(R : J) = (J : J)$. Since R has RTP, $J = \sqrt{J}$. Also, if P is a minimal prime of J, then $(R : P) = (P : P)$ [48, Proposition 2.1], and, as observed above, P is an invertible maximal ideal of $(P : P)$. Thus J is invertible as an ideal of $T := (J : J)$ (Lemma 2.5.14), and, as an ideal of R, it has only finitely many minimal primes, each extending to an invertible maximal ideal of T. It follows that $J = P_1P_2\cdots P_n$ where $P_1, P_2, \ldots, P_n$ are the minimal primes of J.

Further, assume I is a divisorial ideal of R. Then $(I : I) = T$ (Lemma 3.3.5), and $T = J(T : J) = I(R : I)(T : J)$. Hence $IT = I$ is an invertible ideal of T. As J is invertible as an ideal of T, there are invertible ideals B and C of R such that $I = BT$ and $J = CT$. Since $B+C$ is an invertible ideal of T with $J = (B+C)T$, we may assume $B \subseteq C$. From this we have $I(R : B) = T = J(R : C)$ which yields $I = B(R : C)J = AP_1P_2\cdots P_n$ with $A = B(R : C)$ an invertible (integral) ideal R.

To see that (iii) implies (i), assume that $\mathrm{Div}(R) = \{BP_1P_2\cdots P_n \mid B \in \mathrm{Inv}(R)$ and $P_1, P_2, \ldots, P_n$ pairwise comaximal primes of $R\}$. Then each nonzero prime is divisorial. In particular, as R is a Prüfer domain, each maximal ideal is invertible.

Let P be a nonzero nonmaximal prime of R and let $I := r(R : P)$ where $r \in P$ is nonzero. Since P is divisorial and not maximal $(R : P) = (P : P) \supsetneq R$.

Thus I is a proper divisorial ideal of R that is contained in P. Factor $I = BP_1P_2\cdots P_n$ with B invertible and the P_i pairwise comaximal primes of R. Since I is both an ideal of R and an invertible ideal of the proper overring $(R : P)$, I cannot be invertible as an ideal of R. Hence at least one P_i is not a maximal ideal of R, say P_1. On the other hand, $BP_1P_2\cdots P_n$ is an invertible ideal of $(R : P)$. Hence each $P_i(R : P)$ is invertible as an ideal of $(R : P)$, and thus a maximal ideal of $(R : P)$. Note that each maximal ideal of R that does not contain P extends to a maximal ideal of $(R : P)$. Since each prime of $(R : P)$ is extended from a prime of R, all other maximal ideals of $(R : P)$ are extended from primes of R that contain P. It follows that P_1 contains P. If P_1 properly contains P, then $P = P_1P$. But then we have $((R : P) : P_1) = (R : P_1P) = (R : P)$ which implies $P_1(R : P)$ is not invertible as an ideal of $(R : P)$. Thus it must be that $P = P_1$ is a maximal invertible ideal of $(R : P)$. Hence there is a finitely generated ideal $J \subseteq P$ such that $P(R : P) = J(R : P)$ with $\sqrt{J} = P$ and $JR_P = PR_P$. By Theorem 3.3.2, R is a generalized Dedekind domain. □

For more on generalized Dedekind domains, see Gabelli's survey article [29].

3.4 Constructing Almost Dedekind Domains

The purpose of this section is to construct almost Dedekind domains of various sharp and dull degrees. Some of the results in this section are stated for one-dimensional Prüfer domains while others are specific to almost Dedekind domains. Recall from above that $\mathrm{Max}^{\#}(R)$ denotes the set of sharp maximal ideals of a Prüfer domain R (see Page 45), and (usually used only when R is one-dimensional) $\mathrm{Max}^{\dagger}(R) = \mathrm{Max}(R)\backslash\mathrm{Max}^{\#}(R)$ denotes the set of dull maximal ideals of R (see Page 46). In the one-dimensional case, we recursively defined a chain of overrings of R as follows:

$$R_1 := R, \quad R_n := \bigcap\{(R_{n-1}))_N \mid N \in \mathrm{Max}^{\dagger}(R_{(n-1)})\} \text{ for } n \geq 2,$$

stopping at R_m in the event either $R_{m+1} = K$ or $R_m = R_{m+1}(\neq K)$. If $R_m \neq K$, then $R_{m+1} = K$ if and only if each maximal ideal of R_m is sharp. In this case R is said to have sharp degree m. For the case $R = R_1 \subsetneq R_2 \subsetneq R_3 \subsetneq \cdots \subsetneq R_m = R_{m+1}(\neq K)$, then R has dull degree m. In particular, R has dull degree one if and only if each of its maximal ideals is dull. In addition, a fractional ideal I of R has sharp degree n if $IR_n \neq R_n$ but $IR_{n+1} = R_{n+1}$. Note that we will assume the notation for R_n as $\bigcap\{(R_{n-1})_M \mid M \in \mathrm{Max}^{\dagger}(R_{n-1})\}$ to be standard throughout this section.

We start with two examples that make use of a modification of the construction in [58]. Both are based on the first example in [58] which is Example 3.4.13 below. The first is an almost Dedekind domain with a single maximal ideal that is not sharp, and even though $\rho(I)$ (see Page 3.1) is bounded for each nonzero ideal I, the domain does not have radical factorization. In the second, R is an almost Dedekind domain with a nonzero nonunit element b such that $\rho(b)$ is unbounded.

Example 3.4.1. Example of an almost Dedekind domain with a single maximal ideal that is not sharp, and even though $\rho(I)$ is bounded for each nonzero ideal I, the domain does not have radical factorization.

Let $\{X_1, X_2, \ldots, X_n, \ldots\}$ be a countable set of indeterminates over a field F. Set

$$Y_0 := X_1^2X_2^2X_3^2\cdots,\quad Y_1 := X_2^2X_3^2X_4^2\cdots,\quad \ldots,\ Y_n := \prod_{i=n+1}^{\infty} X_i^2,\ \ldots.$$

$D_0 := F[Y_0]_{Y_0}$,
$D_1 := F[X_1, Y_1]_{(X_1)} \bigcap F[X_1, Y_1]_{(Y_1)}$,
$D_2 := F[X_1, X_2, Y_2]_{(X_1)} \bigcap F[X_1, X_2, Y_2]_{(X_2)} \bigcap F[X_1, X_2, Y_2]_{(Y_2)}$,
$D_n := F[X_1, X_2, \ldots, X_n, Y_n]_{(Y_n)} \bigcap (\bigcap_{i=1}^{n} F[X_1, X_2, \ldots, X_n, Y_n]_{(X_i)})$.

Each D_n is a semilocal Dedekind domain with $n+1$ maximal ideals: $P_n := Y_nD_n$ and $N_{n,k} := X_kD_n$ for $1 \le k \le n$.

For P_n, we have $(Y_jD_n)_{P_n} = (P_nD_n)_{P_n}$ for $0 \le j \le n$.

For $N_{n,k}$, we have

$$\begin{aligned}(Y_jD_n)_{N_{n,k}} &= (N_{n,k}D_n)^k_{N_{n,k}}, \text{ for } 0 \le \mathrm{j} < \mathrm{k}, \quad \text{while}\\ (N_{m,k}D_n)_{N_{n,k}} &= (X_kD_n)_{N_{n,k}} = (N_{n,k}D_n)_{N_{n,k}}, \text{ for } \mathrm{k} \le \mathrm{m} \le \mathrm{n}.\end{aligned}$$

Let $R := \bigcup D_n$.

(1) Let I be a nonzero ideal of R. Then there is a D_n such that $I \bigcap D_n$ is a nonzero ideal of D_n. Clearly, each nonzero ideal of D_n contracts to a nonzero ideal of D_0. In particular, $I \bigcap D_0$ is a nonzero ideal of D_0.
(2) The quotient field of R is $K := F(Y_0, X_1, X_2, X_3, \ldots)$.
(3) R is an almost Dedekind domain (that is not Dedekind).

Let Q be a nonzero prime ideal of R. Then $Q \bigcap D_n$ is a nonzero prime ideal of D_n for each n. It follows that $Q \bigcap D_n$ is principal. Let r/s be a nonzero element of QR_Q with $s \in R\backslash Q$. For some i, both r and s are in D_i. Clearly, $r \in Q_i = Q \bigcap D_i$ and $s \in D_i\backslash Q_i$.

Case 1. Some X_k is in Q.

In this case, $r = X_k^q d$ for some positive integer q and some $d \in D_k\backslash Q_k$. It follows that $Q = X_kR$ is a principal prime ideal of R.

Case 2. No X_k is in Q.

In this case, $Q_i = Y_iD_i$ for each i. It follows that Q contains each Y_i. Since $s \in D_i\backslash Q_i$, there is a positive integer q and an element $d \in D_i\backslash Q_i$ such that $r = dY_i^q$. It follows that $Y_iR_Q = QR_Q$ and thus $QR_Q = Y_kR_Q$ for each k.

Since $X_iD_k + Y_kD_k = D_k$ for $i \le k$, R is a one-dimensional domain such that R_Q is principal for each maximal ideal Q. Thus R is an almost Dedekind domain. •

(4) For each $k \ge 1$, let $M_k := X_kR$. Also, let M be the ideal of R generated by the set $\{Y_i \mid 0 \le i < \infty\}$. Then $\mathrm{Max}(R) = \{M, M_1, M_2, M_3, \ldots\}$.

(5) For a monomial g in $\bigcup_n F[X_1, X_2, \ldots, X_n, Y_n]$, there is a smallest n such that $g = bY_n^{r_0}X_1^{r_1}X_2^{r_2}\cdots X_n^{r_n} \in F[X_1, X_2, \ldots, X_n, Y_n]$ with $r_0 \geq 0$. In the case $r_0 \geq 1$, g factors further as $Y_m^{r_0}X_{n+1}^{2r_0}\cdots X_m^{2r_0}X_1^{r_1}X_2^{r_2}\cdots X_n^{r_n}$ in $F[X_1, X_2, \ldots, X_m, Y_m]$ for all $m > n$. But these are the only other factorizations of g. It follows that $\rho(b)$ is bounded for each nonzero nonunit b of R. For a nonzero ideal I, it is clear that $\rho(I) \leq \rho(b)$ for each nonzero $b \in I$. Hence $\rho(I)$ is bounded.
(6) Consider the principal ideal $I := Y_0R$. Clearly, this is contained in each maximal ideal. So $\sqrt{Y_0R}$ is the Jacobson radical of R. Also $Y_0 \in M_k^2$ for all $k \geq 1$ while $Y_0R_M = MR_M$. Hence M^2 does not contain I. On the other hand, since M is the only maximal ideal of R that is not sharp, $\bigcap_k M_k = \sqrt{Y_0R}$. Thus M contains I_2. That R does not have radical factorization follows from Theorem 3.1.8.

Example 3.4.2. Example of an almost Dedekind domain R with a nonzero nonunit b such that $\rho(b)$ is unbounded. As in Example 3.4.1, R has a single maximal ideal that is not sharp. By Theorem 3.1.8, R does not have radical factorization.

Let $\{X_1, X_2, \ldots, X_n, \ldots\}$ be a countable set of indeterminates over a field F. Start with $Y_0 := \prod_{i=1}^{\infty} X_i^i$, and then for $k \geq 1$: $Y_k := \prod_{i=k+1}^{\infty} X_i^i$. Set

$D_0 := F[Y_0]_{Y_0}$,
$D_1 := F[X_1, Y_1]_{(X_1)} \bigcap F[X_1, Y_2]_{(Y_1)}$,
$D_2 := F[X_1, X_2, Y_2]_{(X_1)} \bigcap F[X_1, X_2, Y_2]_{(X_2)} \bigcap F[X_1, X_2, Y_2]_{(Y_2)}$,
$D_n := F[X_1, X_2, \ldots, X_n, Y_n]_{(Y_n)} \bigcap (\bigcap_i F[X_1, X_2, \ldots, X_n, Y_n]_{(X_i)})$.

Each D_n is a semilocal Dedekind domain. The maximal ideals are $N_{n,k} := X_kD_n$ for $1 \leq k \leq n$ and $P_n := Y_nD_n$.

For P_n, we have $(Y_jD_n)_{P_n} = (P_nD_n)_{P_n}$, for $0 \leq j \leq n$.

For $N_{n,k}$, we have

$$(Y_jD_n)_{N_{n,k}} = (N_{n,k}^kD_n)_{N_{n,k}}, \text{ for } 0 \leq j < k, \quad \text{while}$$
$$(N_{m,k}D_n)_{N_{n,k}} = (X_kD_n)_{N_{n,k}} = (N_{n,k}D_n)_{N_{n,k}}, \quad \text{for } k \leq m \leq n.$$

Let $R := \bigcup D_n$.

(1) Let I be a nonzero ideal of R. Then there is a D_n such that $I \bigcap D_n$ is a nonzero ideal of D_n. Clearly, each nonzero ideal of D_n contracts to a nonzero ideal of D_0. In particular, $I \bigcap D_0$ is a nonzero ideal of D_0.
(2) The quotient field of R is $K := F(Y_0, X_1, X_2, X_3, \ldots)$.
(3) R is an almost Dedekind domain (that is not Dedekind).

Let Q be a nonzero prime ideal of R. Then $Q \bigcap D_n$ is a nonzero prime ideal of D_n for each n. It follows that $Q \bigcap D_n$ is principal. Let r/s be a nonzero element of QR_Q with $s \in R\backslash Q$. For some i, both r and s are in D_i. Clearly, $r \in Q_i := Q \bigcap D_i$ and $s \in D_i \backslash Q_i$.

Case 1. Some X_k is in Q.

In this case, $r = dX_k^q$ for some positive integer q and some $d \in D_k \backslash Q_k$. It follows that $Q = X_kR$ is a principal prime ideal of R.

Case 2. No X_k is in Q.

In this case, $Q_i = Y_i D_i$ for each i. It follows that Q contains each Y_i. Since $s \in D_i \backslash Q_i$, there is a positive integer q and an element $d \in D_i \backslash Q_i$ such that $r = dY_i^q$. It follows that $Y_i R_Q = QR_Q$ and thus $QR_Q = Y_k R_Q$ for each k.

Since $X_i D_k + Y_k D_k = D_k$ for $i \leq k$, R is a one-dimensional domain such that R_Q is principal for each maximal ideal Q. Thus R is an almost Dedekind domain. The maximal ideals of R are the principal ideals $M_i := X_i R$ and the unique noninvertible maximal ideal M generated by the set $\{Y_i \mid 0 \leq i\}$.

Finally, the element Y_0 is such that $Y_0 \in M_k^k$ for each $k \geq 1$. Hence $\rho(Y_0)$ is unbounded.

We describe next an alternate approach to building examples of almost Dedekind domains. It is related to (but not the same as) the construction used in [58, Sect. 3].

Example 3.4.3. Example of an almost Dedekind domain with an explicit description of the complete set of its valuation overrings.

Start with $D := F[\mathscr{Y}]$ and $K := F(\mathscr{Y})$ where $\mathscr{Y} := \{Y_i\}_{i=0}^{\infty}$ is a countably infinite set of algebraically independent indeterminates over the field F. Also, let $D_n := F[Y_0, Y_1, \ldots, Y_n]$ and $K_n := F(Y_0, Y_1, \ldots, Y_n)$ for each n.

Set $\mathscr{P}_0 := \{A_{0,1}\}$ where $A_{0,1} := \mathbb{N}$ and set $n_0 := 1$. Then, recursively, for each positive integer m, let $\mathscr{P}_m := \{A_{m,1}, A_{m,2}, \ldots, A_{m,n_m}\}$ be a partition of $\mathbb{N}$ such that $n_m > n_{m-1}$ and, for each $1 \leq i \leq n_m$, $A_{m,i} \subseteq A_{m-1,j}$ for some $A_{m-1,j} \in \mathscr{P}_{m-1}$. The set $\mathscr{P} := \bigcup_m \mathscr{P}_m$ is a countable subset of the power set of $\mathbb{N}$. Let $\{B_i\}_{i=0}^{\infty}$ be the collection of sets in $\mathscr{P}$ ordered in such a way that, if $B_i \in \mathscr{P}_m \backslash \mathscr{P}_{m-1}$ and $B_j \in \mathscr{P}_k$ for some $k < m$, then $j < i$. In particular, we have $j < i$ if $B_i \subsetneq B_j$. Thus $B_0 = \mathbb{N} = A_{0,1}$ and, for $1 \leq i \leq n_1$, $B_i = A_{1,k}$ for some $1 \leq k \leq n_1$. For each positive integer k, let q_k denote the number of sets in $\mathscr{P}_k$ that are not in $\mathscr{P}_{k-1}$. Since $\mathscr{P}_k$ is a proper refinement of $\mathscr{P}_{k-1}$, q_k is positive. We have a strictly increasing sequence of integers $m_0 := 0 < m_1 < m_2 < \cdots$, where $m_k := q_1 + q_2 + \cdots + q_k$ for each positive integer k. Thus $m_k = m_{k-1} + q_k$ for each positive integer k.

Next let $\mathscr{A} := \{A_{m,i_m}\}_{m=0}^{\infty}$ be a family of sets such that $A_{m+1,i_{m+1}} \subseteq A_{m,i_m}$ for each m. We say that $\mathscr{A}$ is a *chain through* $\mathscr{P}$. Let $\mathscr{I}$ be an index set for the collection of all chains through $\mathscr{P}$. Note that $A_{0,1} = B_0$ is in each $\mathscr{A}$. For each $\alpha \in \mathscr{I}$ define a valuation v_α on K as follows:

(i) $v_\alpha(b) = 0$, for all nonzero $b \in F$,
(ii) $v_\alpha(Y_i) = 0$, if $B_i \notin \mathscr{A}_\alpha$,
(iii) $v_\alpha(Y_i) = 1$, if $B_i \in \mathscr{A}_\alpha$,
(iv) extend in the necessary way to products of the indeterminates, then to sums using "min" and finally to quotients.

Let V_α denote the corresponding valuation domain of K, for each $\alpha \in \mathscr{I}$. Each such valuation domain is discrete of rank one.

Since $n_1 > n_0 = 1$, the sets $B_1 \neq B_2$ are in $\mathscr{P}_1$, so at most one of these is in a particular $\mathscr{A}_\alpha$. Thus the element $Y_1 + Y_2$ is a unit in each V_α.

Let $R := \bigcap_{\alpha \in \mathscr{I}} V_\alpha$. We will show that R is an almost Dedekind domain and that $\{V_\alpha\}_{\alpha \in \mathscr{I}}$ is the complete set of valuation overrings of R. Note that since Y_0 is a nonunit of each V_α, it is contained in each maximal ideal of R (for if $\mathrm{Y}_0 R + wR = R$ for some $w \in R$, then w is a unit of each V_α and thus a unit of R).

For each positive integer k, set $g_k := \mathrm{X}^k + \mathrm{Y}_1 + \mathrm{Y}_2$. It is clear that g_2 and g_3 are relatively prime in $K[\mathrm{X}]$. To see that R is a Bézout domain, it suffices to show that both g_2 and g_3 are unit valued in each V_α [57, Corollary 2.7]. It is in fact the case that each g_k is unit valued in each V_α.

Fix a $k \geq 2$ and an $\alpha \in \mathscr{I}$. Since $\mathrm{Y}_1 + \mathrm{Y}_2$ is a unit in V_α, $g_k(t) = t^k + \mathrm{Y}_1 + \mathrm{Y}_2$ is unit for each nonunit $t \in V_\alpha$. Suppose $u \in V_\alpha$ is a unit. Then there is a pair of nonzero polynomials $r, s \in D$ such that $u = r/s$, necessarily with $\mathrm{v}_\alpha(r) = \mathrm{v}_\alpha(s)$. Let r_α and s_α denote the corresponding sums of the monomials of r and s, respectively, with minimum value under v_α. Then r_α^k is the sum of the monomial terms of r^k with minimum value under v_α and s_α^k is the sum of the monomial terms of s^k with minimum value under v_α. We have $\mathrm{v}_\alpha(r^k) = \mathrm{v}_\alpha(r_\alpha^k) = k\mathrm{v}_\alpha(r_\alpha) = k\mathrm{v}_\alpha(s_\alpha) = \mathrm{v}_\alpha(s_\alpha^k) = \mathrm{v}_\alpha(s^k)$. Also both $\mathrm{v}_\alpha(r^k - r_\alpha^k)$ and $\mathrm{v}_\alpha(s^k - s_\alpha^k)$ are strictly larger than $\mathrm{v}_\alpha(r^k) = \mathrm{v}_\alpha(s^k)$. It is clear that $r_\alpha^k + s_\alpha^k(\mathrm{Y}_1 + \mathrm{Y}_2)$ is not the zero polynomial, nor is either of $r_\alpha^k + s_\alpha^k \mathrm{Y}_1$ or $r_\alpha^k + s_\alpha^k \mathrm{Y}_2$. Hence $v_\alpha(r^k + s^k(\mathrm{Y}_1 + \mathrm{Y}_2)) = \mathrm{v}_\alpha(r_\alpha^k + s_\alpha^k(\mathrm{Y}_1 + \mathrm{Y}_2)) = \mathrm{v}_\alpha(s_\alpha^k) = \mathrm{v}_\alpha(s^k)$. It follows that $\mathrm{v}_\alpha(g_k(u)) = 0$ and therefore g_k is unit valued in V_α.

Next, we show that R is an almost Dedekind domain. For each $n \geq 0$ and each $\alpha \in \mathscr{I}$, let $R_n := R \bigcap K_n$ and $V_{\alpha,n} := V_\alpha \cap K_n$. It is clear that $R_n = \bigcap_{\alpha \in \mathscr{I}} V_{\alpha,n}$. Moreover, note that each $V_{\alpha,n}$ is a discrete rank one valuation domain, which is entirely determined by the values $\mathrm{v}_\alpha(\mathrm{Y}_i)$ for $0 \leq i \leq n$. There are only finitely many such valuations and thus R_n is a semilocal Dedekind domain, where each valuation overring has the form $V_{\alpha,n}$ for some $\alpha \in \mathscr{I}$.

Let M be a maximal ideal of R and let $W := R_M$. Since R is Bézout, W is a valuation domain which necessarily contains Y_0 as a nonunit. It follows that, for each positive integer n, $W \bigcap K_n$ is a proper valuation domain of K_n that contains R_n. Hence $W \bigcap K_n = V_{\alpha,n}$ for some $\alpha \in \mathscr{I}$. A consequence is that W is discrete rank one valuation domain. Therefore R is an almost Dedekind domain.

Finally, we show that $W = V_\alpha$ for some $\alpha \in \mathscr{I}$. For each n, there is an $\alpha_n \in \mathscr{I}$ such that $W_n := W \bigcap K_n = V_{\alpha_n,n}$. We consider the sequence of integers $\{m_k\}_{k=0}^\infty$ and the corresponding chain of valuation domains $W_{m_0} \subsetneq W_{m_1} \subsetneq W_{m_2} \subsetneq \cdots$. From above, we know $\mathrm{Y}_0 \in W_{m_0}$ is a nonunit. For each k, W_{m_k} contains D_{m_k}. Hence W_{m_k} contains all Y_is for $0 \leq i \leq m_k$. In the valuation domain W_{m_1}, the element Y_0 is a nonunit as is exactly one other Y_i for some $1 \leq i \leq m_1$; the other Y_js in D_{m_1} are units of W_{m_1}. As k increases, there is at most one integer i between $m_{k-1}+1$ and m_k such that the corresponding Y_i is a nonunit of W_{m_k}. From this analysis, we obtain a (possibly finite) descending chain of sets $B_0 \supsetneq B_{i_1} \supsetneq B_{i_2} \supsetneq \cdots$ corresponding to the Y_is that are nonunits in some W_{m_k}. The set B_{i_1} is a set in the partition $\mathscr{P}_1$. Thus $B_{i_1} = A_{1,j_1}$ for some $1 \leq j_1 \leq n_1$. For each integer k, there is a set $A_{k,j_k} \in \mathscr{P}_k$ that is one of the sets B_{i_h} in the descending chain above. The corresponding family $\mathscr{A} := \{A_{k,j_k}\}$ is a chain through $\mathscr{P}$. Thus there is $\beta \in \mathscr{I}$ such that $\mathscr{A} = \mathscr{A}_\beta$. We have $W_n = V_{\beta,n}$ for each n and therefore $W = V_\beta$.

Next, we apply the techniques introduced in Example 3.4.3 to give an alternate construction of almost Dedekind domains that do not have radical factorization. More precisely, start with $\mathscr{P}_0 := \{\mathbb{N}\}$ and $C_0 := \mathbb{N}$ and then for $m \geq 1$ let $\mathscr{P}_m := \{1, 2, \ldots, m, C_m\}$ where $C_m := \mathbb{N} \backslash \{1, 2, \ldots, m\}$. We number the B_is as follows: $B_i := C_m$ when $i = 2m$ and $B_i := \{m\}$ when $i = 2m - 1$. Thus $B_0 = \mathbb{N}$. The chain $\mathscr{A}_0 := \{C_m\}_{m=0}^{\infty}$ is the only one that is not finite. The others have the form $\mathscr{A}_k := \{C_0, C_1, \ldots, C_{k-1}, \{k\}\}$. In the next two examples, we modify the definitions of the valuation domains associated with the chains through $\mathscr{P}$ to obtain almost Dedekind domains that do not have radical factorization.

Example 3.4.4. Example of an almost Dedekind domain that does not have radical factorization and with an explicit description of the complete set of its valuation overrings.

As in Example 3.4.3, let $D := F[\mathscr{Y}]$ and $K := F(\mathscr{Y})$ where $\mathscr{Y} := \{\mathrm{Y}_i\}_{i=0}^{\infty}$ is a countably infinite set of algebraically independent indeterminates over the field F. For each positive integer n, define a valuation $\mathrm{v}_n : F(\mathrm{Y}_0, \mathrm{Y}_1, \mathrm{Y}_2, \mathrm{Y}_3, \ldots) \backslash \{0\} \to \mathbb{Z}$ by first setting $\mathrm{v}_n(\mathrm{Y}_{2n-1}) = 1$, $\mathrm{v}_n(\mathrm{Y}_{2m}) = 2$ for $0 \leq m \leq n-1$, and $\mathrm{v}_n(b) = \mathrm{v}_n(\mathrm{Y}_k) = 0$ for all $b \in F \backslash \{0\}$ and all $k \notin \{0, 2, 4, \ldots, 2(n-1), 2n-1\}$, extend to products and then to sums using "min", and finally to quotients. Let V_n be the corresponding valuation domain. For $n = 0$, $\mathrm{v}_0(b) = 0$ for all $b \in F \backslash \{0\}$ and for all m, $\mathrm{v}_0(\mathrm{Y}_{2m}) = 1$ and $\mathrm{v}_0(\mathrm{Y}_{2m-1}) = 0$, extend as above. Finally, let $R := \bigcap_{n=0}^{\infty} V_n$ and, for each $n \geq 0$, let M_n be the contraction of the maximal ideal of V_n to R.

(1) R is an almost Dedekind domain such that Y_0 is contained in every maximal ideal.
(2) The set $\{V_n\}_{n=0}^{\infty}$ is the complete set of valuation overrings of R. For each $n \geq 1$, $M_n = \mathrm{Y}_{2n-1}R$ is a maximal ideal of R and $\mathrm{Y}_0, \mathrm{Y}_2, \ldots, \mathrm{Y}_{2n-2} \in M_n^2$. No other Y_is are contained in M_n.
(3) M_0 is not invertible and it is the only maximal ideal of R that is not principal.
(4) Since M_0 is the only noninvertible maximal ideal and every maximal ideal contains Y_0, M_0 contains $\bigcap_{k=1}^{\infty} M_k$. Hence the principal ideal $\mathrm{Y}_0 R$ cannot be factored as a product of radical ideals, and therefore R does not have radical factorization.

Proof. For each V_n, at least one of Y_1 and Y_3 is a unit and Y_0 is a nonunit. As above (Example 3.4.3), $g_k := \mathrm{X}^k + \mathrm{Y}_1 + \mathrm{Y}_3$ is unit valued in V_n for each $k \geq 2$. Hence R is a Bézout domain [57, Corollary 2.7]. Also, Y_0 is contained in each maximal ideal of R.

For each n and m, let $R_n := R \bigcap K_n$ and $V_{m,n} := V_m \bigcap K_n$. Each $V_{m,n}$ is determined by the values of the $\mathrm{v}_m(\mathrm{Y}_i)$s for $0 \leq i \leq n$. The only possible values are 0 and 1 when i is odd, and 0, 1 and 2 when i is even. Hence there are only finitely many distinct $V_{n,m}$. As $R_n = \bigcap_m V_{m,n}$, it is a semilocal Dedekind domain such that each valuation overring is the contraction of some V_m.

Let m be a positive integer. If $2m-1 \leq n$, then $V_{m,n}$ is the only valuation overring of R_n that contains Y_{2m-1} as a nonunit. On the other hand, if $2m-1 > n$, then $V_{m,n}$ contains each $\mathrm{Y}_{2k} \in D_n$ and it follows that $V_{m,n} = V_{0,n}$ for all such m.

Let M be a maximal ideal of R. Then R_M is a valuation domain which contains Y_0 as a nonunit. It follows that $R_M \bigcap K_n$ is a proper valuation overring of R_n for each n. Hence $R_M \bigcap K_n$ is discrete rank one valuation domain. Thus R_M is a discrete rank one valuation domain and therefore R is an almost Dedekind domain. If $Y_{2m-1} \in MR_M$ for some $m \geq 1$, then $R_M \bigcap K_n = V_{m,n}$ for all $n \geq 2m-1$. It follows that $R_M = V_m$ in this case. The only other possibility is that $Y_{2m} \in MR_M$ for each $m \geq 0$. In this case, $R_M = V_0$. Therefore $\{V_m\}_{m=0}^{\infty}$ is the complete set of valuation overrings of R.

We have $Y_0 \in M_n^2$ for each $n \geq 1$, but $Y_0 \in M_0 \backslash M_0^2$. As M_0 is the only maximal ideal that is not principal, it contains $\bigcap_{n=1}^{\infty} M_n$. Hence R does not have radical factorization (Theorem 3.1.8). □

Next, we give another example of an almost Dedekind domain of the type considered in Example 3.4.2.

Example 3.4.5. Example of an almost Dedekind domain with an explicit description of the complete set of its valuation overrings and with a nonzero nonunit element b such that $\rho(b)$ is unbounded. In particular, this domain does not have radical factorization (Theorem 3.1.8).

As in Example 3.4.3, let $D := F[\mathscr{Y}]$ and $K := F(\mathscr{Y})$ where $\mathscr{Y} := \{Y_i\}_{i=0}^{\infty}$ is a countably infinite set of algebraically independent indeterminates over the field F. For each positive integer n, define a valuation $v_n : F(Y_0, Y_1, Y_2, Y_3, \ldots)\backslash\{0\} \to \mathbb{Z}$ by setting $v_n(Y_{2n-1}) = 1$, $v_n(Y_{2m}) = n$ for $0 \leq m \leq n-1$, and $v_n(b) = v_n(Y_k) = 0$ for all $b \in F\backslash\{0\}$ and all $k \notin \{0, 2, 4, \ldots, 2n-2, 2n-1\}$, extend to products and then to sums using "min", and finally to quotients. Let V_n be the corresponding valuation domain. For $n = 0$, set $v_0(Y_{2m}) = 1$ and $v_0(Y_{2m+1}) = 0 = v_0(b)$ for all nonzero $b \in F$ and all $m \geq 0$, and again extend to products and then to sums using "min", and finally to quotients. We again set $R := \bigcap_{n=0}^{\infty} V_n$ and, for each $n \geq 0$, let M_n be the contraction of the maximal ideal of V_n to R.

(1) R is an almost Dedekind domain such that Y_0 is contained in every maximal ideal.
(2) For each $n \geq 1$, $M_n := Y_{2n-1}R$ is a maximal ideal of R and $Y_0 \in M_n^n$. So $\rho(Y)$ is unbounded. Therefore R does not have radical factorization.

For (1), adapt the proof used in the previous example. The statement in (2) is clear.

Lemma 3.4.6. (Loper–Lucas [58, Lemma 2.1]) *Let R be a one-dimensional Prüfer domain.*

(1) If M is a maximal ideal of R_n for some $n \geq 2$, then there is a maximal ideal P of R such that $PR_n = M$ and PR_{n-1} is a dull prime of R_{n-1}.
(2) If $P \in Max(R)$ survives in R_n, then

(a) PR_{n-1} is a dull prime of R_{n-1}, and
(b) $PR_n \in Max^{\#}(R_n)$ if and only if there is a finitely generated ideal I of R such that P is the only maximal ideal of R that both contains I and survives in R_n.

Proof. (1) Let M be a maximal ideal of R_n. Since R is a one-dimensional Prüfer domain, each prime of R_n is extended from a prime of R. Thus $M = PR_n$ for some $P \in \operatorname{Max}(R)$. To show that PR_{n-1} is a dull prime of R_{n-1}, consider what happens to a sharp prime Q of R_{n-1}. Since R_{n-1} is a one-dimensional Prüfer domain [34, Theorem 26.1], Q is the radical of a finitely generated ideal J. Thus J^{-1} is contained in each localization of R_{n-1} at a dull prime. Hence J^{-1} is contained in R_n. But then $JR_n = JJ^{-1}R_n = R_n$ and therefore $QR_n = R_n$. Hence PR_{n-1} must be a dull prime of R_{n-1}.

To prove (2), suppose $P \in \operatorname{Max}(R)$ survives in R_n. Then, by the above, PR_{n-1} must be a dull prime of R_{n-1}. Obviously, if there is a finitely generated ideal I of R such that PR_n is the only maximal ideal of R_n that contains IR_n, then $PR_n = \sqrt{IR_n}$ is a sharp prime of R_n. Conversely, if PR_n is a sharp prime of R_n, then there is a finitely generated ideal J_n of R_n for which $PR_n = \sqrt{J_n}$. Since PR_n is generated by the elements of P, there is a finitely generated ideal I of R whose extension to R_n is contained in PR_n and contains J_n. □

Note that if PR_n is a sharp prime of R_n, any finitely generated ideal I that satisfies the conditions in Lemma 3.4.6(2) must be contained in infinitely many primes which do not survive in R_n, for otherwise P will be a sharp prime of R_k for some $k < n$ and thus not survive in R_n.

It is known that if a finitely generated ideal of an almost Dedekind domain is contained in only finitely many maximal ideals, then the ideal is a product of positive powers of these maximal ideals [34, Theorem 37.5]. The converse is trivial. In our next lemma, we show that the finitely generated fractional ideals of sharp degree one in an almost Dedekind domain are those that can be factored into finite products of nonzero powers of maximal ideals.

Lemma 3.4.7. (cf. also [34, Theorem 37.5]) *Let R be an almost Dedekind domain and let I be a finitely generated fractional ideal of R. Then I is a finite product of nonzero powers of maximal ideals if and only if I has sharp degree one.*

Proof. First, assume $I = M_1^{r_1}M_2^{r_2}\cdots M_n^{r_n}$ with each r_i a nonzero integer and no $M_i^{r_i} = R$. Since I is finitely generated, it is invertible. Thus each M_i is invertible and therefore a sharp prime. We have $M_iR_2 = R_2$ for each i, and the same happens for M_i^{-1}. Thus $IR_2 = R_2$ and we have that I has sharp degree one.

To complete the proof assume I has sharp degree one. Then $IR_2 = R_2$. Partition $\operatorname{Max}(R)$ into sets (possibly with some, but not all, empty)

$$\mathcal{M}^0(I) := \{P \in \operatorname{Max}(R) \mid IR_P = R_P\},$$

$$\mathcal{M}^+(I) := \{P \in \operatorname{Max}(R) \mid IR_P \subseteq PR_P\}, \text{ and}$$

$$\mathcal{M}^-(I) := \{P \in \operatorname{Max}(R) \mid I^{-1}R_P \subseteq PR_P\}.$$

Since each dull prime survives in R_2 and $IR_2 = R_2$, each dull prime must be in the set $\mathcal{M}^0(I)$. Therefore $R_I^+ := \bigcap_{P\in\mathcal{M}^+(I)} R_P$ and $R_I^- := \bigcap_{P\in\mathcal{M}^+(I)} R_P$ are both Dedekind domains with nonzero Jacobson radicals. Thus each is semilocal which means that both $\mathcal{M}^+(I)$ and $\mathcal{M}^-(I)$ are finite sets. Note that $\mathcal{M}^-(I)$ is empty if I is an integral ideal of R, but both may be nonempty if I is fractional. Set $\mathcal{M}^+(I) = \{M_1, M_2, \ldots, M_n\}$ and $\mathcal{M}^-(I) = \{N_1, N_2, \ldots, N_m\}$. It follows that $IR_I^+ = M_1^{r_1} M_2^{r_2} \cdots M_n^{r_n} R_I^+$ and $I^{-1} R_I^- = N_1^{s_1} N_2^{s_2} \cdots N_m^{s_m} R_I^-$ for some positive integers r_i and s_j. We also have $IR_I^- = N_1^{-s_1} N_2^{-s_2} \cdots N_m^{-s_m} R_I^-$. By checking locally, we see that $I = M_1^{r_1} M_2^{r_2} \cdots M_n^{r_n} N_1^{-s_1} N_2^{-s_2} \cdots N_m^{-s_m}$. This representation is unique since each M_i and N_j is a maximal ideal. □

If R is an almost Dedekind domain, then a maximal ideal P is sharp if and only if it is invertible. Hence P has sharp degree n if and only if PR_k is dull for each $k < n$ and PR_n is an invertible maximal ideal of R_n. Also, for $n > 1$, $\mathrm{Max}(R_n) = \{PR_n \mid P \in \mathrm{Max}^\dagger(R_{n-1})\}$ whenever $R_n \neq K$.

Theorem 3.4.8. *Let R be an almost Dedekind domain. For each positive integer k and each prime P_α of sharp degree k, let J_α be a finitely generated ideal of R such that $J_\alpha R_{P_\alpha} = P_\alpha R_{P_\alpha}$ and J_α is contained in no other prime of R_k. If I is a finitely generated fractional ideal of R of finite sharp degree, then I factors uniquely into a finite product of nonzero powers of ideals from the family $\{J_\alpha\}$. In particular, the members of the family $\{J_\alpha\}$ are distinct.*

Proof. First note that, if P_α is a sharp prime of R, then by checking locally we see that the corresponding J_α is simply P_α itself. Moreover, by checking locally in R_k, we see that if P_α has sharp degree k, then $J_\alpha R_k = P_\alpha R_k$. Let P_α and P_β be distinct maximal ideals of R with P_α of finite sharp degree k. Then, in R_k, we have $J_\alpha R_k = P_\alpha R_k$ with $P_\alpha R_k$ a maximal ideal of R_k. Thus the only way to have $P_\beta R_k$ contain J_α is to have P_β blow up in R_k. In such a case P_β would have sharp degree $m < k$. While it might be that $J_\alpha R_{P_\beta} = P_\beta R_{P_\beta}$, $J_\alpha R_m$ would be contained in $P_\alpha R_m$ so that $J_\alpha R_m$ cannot equal $P_\beta R_m$. Thus $J_\alpha \neq J_\beta$. It follows that if both the distinct maximal ideals P_α and P_β have finite sharp degree, then $J_\alpha \neq J_\beta$. Moreover, no nonzero powers can be equal and $J_\alpha R_n = R_n$ for each $n > k$.

We will take care of uniqueness first. For this it suffices to show that there is no nontrivial factorization of R since each of the J_αs is invertible. Assume $R = \prod J_{m,i}^{e_{m,i}}$ is a finite factorization of R over the set $\{J_\alpha\}$ with each $J_{m,i}$ having sharp degree m and $e_{m,i}$ an integer, perhaps 0. Let n denote the highest sharp degree of any "factor" of R. Then, in R_n, we have $R_n = \prod J_{n,i}^{e_{n,i}}$ since $J_{m,i} R_n = R_n$ for $m < n$. As $J_{n,i} R_n = P_{n,i} R_n$ is a maximal ideal of R_n, it must be that each $e_{n,i} = 0$. Thus the factors $J_{n,i}^{e_{n,i}}$ are all superfluous. Continue the process to show all $e_{m,i}$ are 0.

For existence of factorizations we use induction and Lemma 3.4.7.

By Lemma 3.4.7, if I has sharp degree one, then I is a product of nonzero powers of finitely many sharp maximal ideals, say $I = M_1^{e_1} M_2^{e_2} \cdots M_n^{e_n}$.

Now, assume I has sharp degree two. Then IR_2 is a finitely generated fractional ideal of R_2 whose sharp degree as an ideal of R_2 is one. Thus by Lemma 3.4.7, there are finitely many maximal ideals $P_1 R_2, P_2 R_2, \ldots, P_n R_2$ of R_2 which locally contain either IR_2 or $(IR_2)^{-1}$. For each i, we have a finitely generated ideal J_i in the set $\{J_\alpha\}$ such that $J_i R_2 = P_i R_2$. Thus in R_2, we can factor IR_2 uniquely as $P_1^{e_1} P_2^{e_2} \cdots P_n^{e_n} R_2$ for some nonzero integers $e_1, e_2 \ldots, e_n$. This factorization is the same as the factorization $J_1^{e_1} J_2^{e_2} \cdots J_n^{e_n} R_2$ since $P_i R_2 = J_i R_2$ for each i. Let $J := J_1^{e_1} J_2^{e_2} \cdots J_n^{e_n}$. Then $I(R : J)R_2 = R_2$. As both I and $(R : J)$ are finitely generated fractional ideals of R, $I(R : J)$ is a finitely generated fractional ideal of R. It has sharp degree one since $I(R : J)R_2 = R_2$. Thus by Lemma 3.4.7, there are finitely many maximal ideals $M_1, M_2, \ldots, M_m$ such that $I(R : J) = M_1^{r_1} M_2^{r_2} \cdots M_m^{r_m}$ for some nonzero integers r_i. Therefore $I = I(R : J)J = M_1^{r_1} M_2^{r_2} \cdots M_m^{r_m} J_1^{e_1} J_2^{e_2} \cdots J_n^{e_n}$.

Now, assume a factorization exists for each finitely generated fractional ideal of sharp degree k or less (in every almost Dedekind domain). Let I be a finitely generated fractional ideal of R which has sharp degree $k+1$. Then IR_2 is a finitely generated fractional ideal of R_2 which has sharp degree k. Thus IR_2 factors into a finite product, say $IR_2 = J_1^{e_1} J_2^{e_2} \cdots J_m^{e_m} R_2$. To complete the proof simply repeat the steps used above for the case of an ideal of sharp degree 2. Namely, set $J := J_1^{e_1} J_2^{e_2} \cdots J_m^{e_m}$ and factor the fractional ideal $I(R : J)$ over the sharp primes of R. This establishes existence of a factorization. □

One special case we wish to consider is the one of an almost Dedekind domain with exactly one dull prime.

Theorem 3.4.9. *Let R be a one-dimensional Prüfer domain. Then R is an almost Dedekind domain with at most one noninvertible maximal ideal if and only if there is an element $d \in R$ such that, for each finitely generated nonzero ideal I, there is a finite set of maximal ideals $\{M_1, M_2, \ldots, M_m\}$ and integers $e_1, e_2, \ldots, e_m$ and n with $n \geq 0$ such that $I = M_1^{e_1} M_2^{e_2} \cdots M_m^{e_m} (d)^n$. Moreover, if either (hence both) holds and R is not Dedekind, then the element $d \in R$ must be such that $dR_P = PR_P$ for the noninvertible maximal ideal P of R and the set $\{dR\} \bigcup Max^{\#}(R)$ is a factoring set for R such that each finitely generated fractional ideal factors uniquely.*

Proof. For R Dedekind, we simply set $d = 1$. Thus we may assume R is not Dedekind.

Assume R is an almost Dedekind domain with one noninvertible maximal ideal P. Then $R_2 = R_P$ and therefore there is an element $d \in R$ such $PR_2 = dR_2$ since R_P is a DVR. Thus by Theorem 3.4.8, the set $\{dR\} \bigcup \text{Max}^{\#}(R)$ is a factoring set for R such that each finitely generated fractional ideal factors uniquely as a finite product of nonzero powers of members of this set.

For the converse, assume there is an element $d \in R$ such that each finitely generated nonzero ideal can be written in the form $M_1^{e_1} M_2^{e_2} \ldots M_m^{e_m} (d)^n$ where each M_i is a maximal ideal, each e_i is a nonzero integer and n is a non-negative integer. Let I be a finitely generated ideal of R and write $I = M_1^{e_1} M_2^{e_2} \cdots M_m^{e_m} (d)^n$ with no $M_i^{e_i} = R$. Since R is a Prüfer domain, I is invertible. Combining this with the assumption that $M_i^{e_i}$ is not equal to R, we have that each M_i is invertible.

As we are not assuming that R is almost Dedekind, we need to show that each sharp prime is invertible. Let $M \in \text{Max}(R)$ be a noninvertible maximal ideal of R, such a maximal ideal exists since we are assuming R is not Dedekind. Then no (nonzero) power of M can appear as a nontrivial factor (i.e., not R) in a factorization of a finitely generated ideal. Hence d must be contained in M and each finitely generated ideal contained in M must have a positive power of (d) in a factorization. It follows that $MR_M = dR_M$ and R_M is a DVR. Such a prime M cannot be sharp, since to be sharp it would have to contain a finitely generated ideal J that is contained in no other maximal ideal of R. By checking locally, we would then find that M is the finitely generated (and therefore invertible) ideal $dR + J$. So all of the sharp primes are invertible and the dull ones are locally principal. Hence R is an almost Dedekind domain.

We next show that R has at most one dull maximal ideal. By way of contradiction, assume P_1 and P_2 are distinct dull maximal ideals of R. Let b be an element of P_1 that is not in P_2 and write $(b) = M_1^{e_1} M_2^{e_2} \cdots M_m^{e_m} (d)^n$ with no $M_i^{e_i} = R$. As above, each M_i must be invertible. Thus neither P_1 nor P_2 appears in the factorization. Therefore n must be positive and d must be an element of P_1. By repeating this argument for an element in P_2 that is not in P_1, we find that d is also in P_2. But then we have $(b)R_{P_2} = (d)^n R_{P_2} \subseteq P_2 R_{P_2}$ which is a contradiction. Hence there must be exactly one dull maximal ideal and the rest are both sharp and invertible. □

Next we give a general construction scheme for producing an almost Dedekind domain which will have a factoring family for finitely generated ideals. By carefully selecting the members, we can produce a family such that each nonzero finitely generated fractional ideal will factor uniquely over the underlying set of allowable factors.

Theorem 3.4.10. *Let $D_1 \subsetneq D_2 \subsetneq \cdots$ be a chain of Dedekind domains which satisfy all of the following*

(a) *For $i < j$, each maximal ideal of D_i survives in D_j.*
(b) *Each maximal ideal of D_j contracts to a maximal ideal of D_1.*
(c) *If M' is a maximal ideal of D_j and $M := M' \bigcap D_1$, then $M(D_j)_{M'} = M'(D_j)_{M'}$.*

Let $R := \bigcup D_n$. Then the following hold.

(1) *R is an almost Dedekind domain.*
(2) *For $i < j$, each maximal ideal of D_i is contained in only finitely many maximal ideals of D_j. Moreover, if M_i is a maximal ideal of D_i and $M_{j,1}, M_{j,2}, \ldots, M_{j,r}$ are the maximal ideals of D_j that contain M_i, then $M_i D_j = \prod M_{j,k}$.*
(3) *For each finitely generated ideal I of R, there is a finitely generated ideal I_i of some D_i such that $I = I_i R$.*
(4) *A maximal ideal M is a sharp prime of R if and only if $M = M_n R$ for some $M_n := M \bigcap D_n$.*
(5) *There is a family $\{J_\alpha\}$ that is a factoring family for R for which each nonzero finitely generated fractional ideal can be factored uniquely over the underlying set of the family.*
(6) *R is a Dedekind domain if and only if each maximal ideal of D_1 is contained in only finitely many maximal ideals of R.*

Proof. For each n, we let K_n denote the quotient field of D_n.

(1) Let M be a maximal ideal of R and let $M_i := M \bigcap D_i$. Obviously, some M_i is not zero. But then no M_i is zero. Let $r/s \in MR_M$ with $s \in R\backslash M$. For some i, both r and s are in D_i. So $r \in M_i$. But then there is an element $b \in M_1$ and an element $t \in D_i\backslash M_i$ such that $b/t = r/s$. It follows that $MR_M = M_i R_M$ for each i. Since each D_i is Dedekind, $M_i(D_i)_{M_i}$ is principal. Thus MR_M is principal and height one. Hence R is an almost Dedekind domain.

(2) The first statement is a simple consequence of the fact that each ideal of a Dedekind domain is contained in only finitely many maximal ideals. For the second, let M_i be a maximal ideal of D_i and let $M_{j,1}, M_{j,2}, \ldots, M_{j,r}$ be the maximal ideals of D_j that contain M_i. Since the $M_{j,k}$s are maximal ideals of D_j, their intersection is the same as their product. Thus $M_i D_j$ is contained in $\prod M_{j,k}$. Equality comes from our assumption (c); i.e., $M_i(D_j)_{M_{j,k}} = M_{j,k}(D_j)_{M_{j,k}}$.

(3) Since the set $\{D_i\}$ forms a chain, each finitely generated ideal of R can be generated by some finite subset of some D_i.

(4) Since R is an almost Dedekind domain, a maximal ideal is sharp if and only if it is finitely generated. Hence by (3), M is sharp if and only if some D_n contains a generating set for M. As $M \bigcap D_n = M_n$ is a maximal ideal of D_n, then $M = M_n R$.

(5) For each maximal ideal M of R and each positive integer i, let $M_i := M \bigcap D_i$. It is easy to see that $M = \bigcup M_i$. Hence the chain $\{M_i\}$ is uniquely determined by M. Moreover, if $N_1 \subseteq N_2 \subseteq \cdots$ is a chain with each N_k a maximal ideal of D_k, then $N := \bigcup N_i$ is a maximal ideal of R. We say that $\{M_i\}$ is the chain determined by M, and that N is the maximal ideal determined by the chain $\{N_i\}$. Each member N_j of the chain $\{N_i\}$ uniquely determines the members of the chain below it since we have $N_i = N_j \bigcap D_i$ for each $i < j$. Thus for each j, N is determined by the truncated chain $\{N_i\}_{i=j}^{\infty}$.

Since each D_n is a Dedekind domain, the primes of any ring between D_n and its quotient field, K_n, are all extended from primes of D_n. With the restrictions we have placed on the maximal ideals, the quotient field of D_n (for $n \geq 2$) properly contains the quotient field of D_{n-1} with $D_{n-1} = D_n \bigcap K_{n-1}$.

Let I be a fractional ideal of D_{n-1}, for $n \geq 2$. We will show that $I = ID_n \bigcap K_{n-1}$. We at least have $I \subseteq ID_n \bigcap K_{n-1}$. Since D_{n-1} is a Dedekind domain, each of its fractional ideals is invertible and therefore divisorial. Thus it suffices to show that each element of $(D_{n-1} : I)$ multiplies $ID_n \bigcap K_{n-1}$ into D_{n-1}. Since both $(D_{n-1} : I)$ and $ID_n \bigcap K_{n-1}$ are contained in K_{n-1}, the product is there as well. Now, use the fact that both I and $ID_n \bigcap K_{n-1}$ will generate ID_n together with the fact that each element of $(D_{n-1} : I)$ is in $(D_n : ID_n)$ to verify that $(D_{n-1} : I)(ID_n \bigcap K_{n-1})$ is contained in D_{n-1}. Thus $ID_n \bigcap K_{n-1} = I$.

For each n and each maximal ideal M_n of D_n, let $\mathscr{C}(M_n)$ denote *the set of maximal ideals of D_{n+1} that contract to M_n*. The set $\mathscr{C}(M_n)$ is finite since D_{n+1} is a Dedekind domain. Now, select a member M_{n+1} of $\mathscr{C}(M_n)$ and then set $\mathscr{F}(M_n) := \mathscr{C}(M_n)\backslash\{M_{n+1}\}$. We will refer to M_{n+1} as a (or the) *discarded prime* sometimes including the phrase "of D_{n+1}" for emphasis. We refer to the members of $\mathscr{C}(M_n)$ as *conjugates* or *conjugate factors* of M_n. If $\mathscr{C}(M_n)$ is a singleton set, then $M_n D_{n+1}$ is a maximal ideal of D_{n+1} and $\mathscr{F}(M_n)$ will be the empty set. Note that, in this case, we will refer to $M_n D_{n+1}$ as a discarded prime (of D_{n+1}) even if M_n is not a discarded prime of D_n. For $n \geq 1$, let $\mathscr{F}(D_n) := \bigcup\{\mathscr{F}(M_n) \mid M_n \in \mathrm{Max}(D_n)\}$, then set $\mathscr{F}(R) := \bigcup \mathscr{F}(D_n)$. Next, let $\mathscr{G}(R) := \{MR \mid M \in \mathrm{Max}(D_1) \bigcup \mathscr{F}(R)\}$. We will show that each finitely generated ideal of R can be factored uniquely as a finite product of integer powers of ideals from the set $\mathscr{G}(R)$. Then we will show how to build a factoring family for R using only the members of $\mathscr{G}(R)$.

For each positive integer n, let $\mathscr{G}(D_n)$ denote the set $\{PD_n \mid P \in \mathrm{Max}(D_1)$ or $P \in \mathscr{F}(D_k)$ for some $1 \leq k < n\}$. We use induction to show that each nonzero fractional ideal of D_n can be factored uniquely as a finite product of nonzero integer powers of members of $\mathscr{G}(D_n)$. Since $IR \bigcap K_n = I$ for each fractional ideal I of D_n, each finitely generated fractional ideal of R will factor uniquely over $\mathscr{G}(R)$.

Let I_n be a nonzero fractional ideal of D_n. The result is trivial if $n = 1$ since $\mathscr{G}(D_1) = \mathrm{Max}(D_1)$, so we move on to the case $n = 2$. Since D_2 is a Dedekind domain, each nonzero fractional ideal has sharp degree one. Thus Lemma 3.4.7 guarantees that the fractional ideal I_2 factors uniquely as a finite product of nonzero integer powers of maximal ideals of D_2, say $I_2 = \prod_{i=1}^{k} P_i^{r_i}$, with $P_i \in \mathrm{Max}(D_2)$. If each P_i is in $\mathscr{F}(D_1)$, then we at least have existence of a factorization. If not, then some P_i must be a discarded prime. In such a case, there is a maximal ideal M_i of D_1 that has P_i as a factor in D_2. If P_i is the only maximal ideal of D_2 that is a factor of M_i, then we have $M_i D_2 = P_i$, and we simply "substitute" $M_i D_2$ for P_i (they are in fact equal). On the other hand, if M_i has more than one prime factor in D_2, then the other factors are in the set $\mathscr{G}(D_2)$ as only one prime factor is discarded from a set of conjugates. In this case, $M_i D_2 = P_i Q_1 Q_2 \cdots Q_m$ where the Q_ℓs are the conjugates of P_i each of which is in $\mathscr{G}(D_2)$. Thus $P_i = M_i D_2 \prod_{\ell=1}^{m} Q_\ell^{-1}$ and therefore $P_i^{r_i}$ can be replaced by the product $M_i^{r_i} D_2 \prod_{\ell=1}^{m} Q_\ell^{-r_i}$. By doing this for each of the discarded primes in the product $\prod P_i^{r_i}$ we obtain a finite factorization of I_2 using ideals in the set $\mathscr{G}(D_2)$.

Now assume that for each $1 \leq k < n$, each finitely generated ideal of D_k can be factored into a finite product of nonzero integer powers of members of the set $\mathscr{G}(D_k)$. Let I_n be a nonzero fractional ideal of D_n. As above, D_n is a Dedekind domain so I_n factors uniquely as finite product of nonzero powers of maximal ideals P_is of D_n. If each P_i is in $\mathscr{F}(D_{n-1})$, then we have a factorization of I_n over $\mathscr{G}(D_n)$. If not, then some P_i must be a discarded prime. Let $Q_i := P_i \bigcap D_{n-1}$ and let $\prod_{a=1}^{b} N_a^{s_a} D_{n-1}$ be a factorization over the set $\mathscr{G}(D_{n-1})$ for Q_i. If $Q_i D_n = P_i$, we simply take the factorization of Q_i in D_{n-1} and extend each factor to D_n to get a replacement for P_i. If $Q_i D_n \neq P_i$, then $Q_i D_n = P_i \prod_{c=1}^{s} M_c$ where the M_cs are the conjugates to P_i.

Thus each is in the set $\mathscr{G}(D_n)$. As in the case $n = 2$, $P_i = Q_i D_n \prod_{c=1}^{s} M_c^{-1}$. Now, replace $Q_i D_n$ by $\prod_{a=1}^{b} N_a^{s_a} D_n$ to get $P_i = \prod_{a=1}^{b} N_a^{s_a} D_n \cdot \prod_{c=1}^{s} M_c^{-1}$. Do this for each discarded prime in the original factorization of I_n. This will yield a finite factorization of I_n over the set $\mathscr{G}(D_n)$. Extending both I_n and each factor to R will yield a finite factorization of $I_n R$ over the set $\mathscr{G}(R)$. As each finitely generated ideal of R is the extension of some ideal I_n in some D_n, we have that each finitely generated ideal of R has a finite factorization over the set $\mathscr{G}(R)$.

Since D_1 is a Dedekind domain, Lemma 3.4.7 implies each fractional ideal of D_1 can be factored uniquely over the set $\text{Max}(D_1)$. This forms the base for a proof by induction. Assume that for each integer $1 \le k < n$, each fractional ideal of D_k can be factored uniquely over the set $\mathscr{G}(D_k)$. Since each member of $\mathscr{G}(D_k)$ extends to a member of $\mathscr{G}(D_m)$ for each $m > k$, our assumption is equivalent to simply saying that each fractional ideal of D_{n-1} factors uniquely over $\mathscr{G}(D_{n-1})$.

Let J be a nonzero fractional ideal of D_n and let $J = \prod_{i=1}^{m} Q_i^{r_i} \cdot \prod_{a=1}^{n} (P_a D_n)^{s_a}$ with the Q_is in $\mathscr{F}(D_{n-1})$ and the P_as in $\mathscr{G}(D_n)\backslash\mathscr{F}(D_{n-1})$. Suppose $\prod_{c=1}^{k} N_c^{t_c} \cdot \prod_{e=1}^{q} (M_e D_n)^{u_e}$ with the N_cs in $\mathscr{F}(D_{n-1})$ and the M_es in $\mathscr{G}(D_n)\backslash\mathscr{F}(D_{n-1})$ is a potentially different factorization of J over $\mathscr{G}(D_n)$. By multiplying by inverses, we may obtain $\prod_{i=1}^{m} Q_i^{r_i} \cdot \prod_{c=1}^{k} N_c^{-t_c} = \prod_{e=1}^{q} (M_e D_n)^{u_e} \cdot \prod_{a=1}^{n} (P_a D_n)^{-s_a}$. Since the left hand side of the equation is a product of integer powers of maximal ideals of D_n, its form is unique once common factors are combined. Moreover, the primes on the left hand side are all nontrivial factors of primes from D_{n-1} and for each N_c and Q_i exactly one conjugate factor cannot appear in this product. On the other hand, each M_e and each P_a is a prime of some smaller D_k that either factors nontrivially in D_n or generates a maximal ideal of D_n. Those that generate maximal ideals of D_n can have no factor on the left hand side of the equation and those that have a nontrivial factorization must be missing the corresponding discarded prime on the left hand side. Thus the left hand side must reduce to D_n. This can occur only if the factors in $\prod Q_i^{r_i}$ are simply a rearrangement of the factors in $\prod N_c^{t_c}$. As each factor is an invertible fractional ideal of D_n, we may cancel the products $\prod Q_i^{r_i}$ and $\prod N_c^{t_c}$ and obtain $\prod (P_a D_n)^{s_a} = \prod (M_e D_n)^{u_e}$. Since $I D_n \bigcap K_{n-1} = I$ for each fractional ideal of D_{n-1}, we have $\prod (P_a D_{n-1})^{s_a} = \prod (M_e D_{n-1})^{u_e}$. Now simply invoke the induction hypothesis to get uniqueness of factorizations.

It remains to show that we can build a factoring family using only the members of the set $\mathscr{G}(R)$. This is actually relatively easy because given any ideal J in $\mathscr{G}(R)$, there is some unique integer n such that $J = P_n R$ for some maximal ideal P_n of D_n that is not a discarded prime of D_n. This places P_n in $\mathscr{G}(D_n)$. While there may be primes above P_n that are not discarded primes, there is a unique chain of primes $P_{n+1} \subsetneq P_{n+2} \subsetneq \cdots$ with each P_k a discarded prime of D_k and $P_k \bigcap D_n = P_n$. Let P_α be the prime of R

determined by this particular chain through P_n and set $J_\alpha := J := P_n R$. Since $P_n = P_\alpha \bigcap D_n$, $J_\alpha R_{P_\alpha} = P_\alpha R_{P_\alpha}$. Note that this means there is a natural one-to-one correspondence between the set $\mathcal{G}(R)$ and the subset of $\mathrm{Max}(R)$ consisting of those maximal ideals M_β for which there is largest integer n such that $M_\beta \bigcap D_n$ is not a discarded prime. There may be a(or even infinitely many) maximal ideal M_σ of R for which there is no largest integer n such that $M_\sigma \bigcap D_n$ is not a discarded prime. For such a prime, simply set J_σ equal to any member $J := M_n R$ of $\mathcal{G}(R)$ such that $M_n = M_\sigma \bigcap D_n$. With this we have a factoring family for R such that the underlying set allows for unique factorization of nonzero finitely generated fractional ideals.

(6) By the proof of (5), we see that if each maximal ideal of D_1 is contained in only finitely many maximal ideals of R, then each maximal ideal of R is finitely generated. Thus R is a Dedekind domain. Conversely, if R is a Dedekind domain, each maximal ideal of R is finitely generated. Thus for $M \in \mathrm{Max}(R)$, there is a maximal ideal M_n of some D_n such that $M = M_n R$. Assume $M_1 \in \mathrm{Max}(D_1)$ is contained in infinitely many maximal ideals of R. Then there must be a chain of maximal ideals $\{M_n\}$ with each M_n a maximal ideal of D_n such that each M_n is contained in infinitely many maximal ideals of R. Thus none of these ideals can generate a maximal ideal of R. Hence $M = \bigcup M_n$ must be a maximal ideal of R which is not finitely generated, a contradiction of the Dedekind assumption. Therefore each maximal ideal of D_1 is contained in only finitely many maximal ideals of R.

□

The examples that follow make use of ideas in Theorem 3.4.10 and some of the notation and terminology used in Example 3.4.3.

Notation 3.4.11. Let $\mathcal{P}_0 := \{\mathbb{N}\}$ and let $\mathcal{P}_1 := \{A_{1,1}, A_{1,2}, \ldots, A_{1,n_1}\}$ be a partition of $\mathbb{N}$ into finitely many disjoint nonempty sets with $n_1 > 1$. Recursively, for each positive integer $m > 1$, let $\mathcal{P}_m := \{A_{m,1}, A_{m,2}, \ldots, A_{m,n_m}\}$ be a refinement of the partition $\mathcal{P}_{m-1}$ with $n_m > n_{m-1}$ but allowing some $A_{m-1,k}$ to survive intact in $\mathcal{P}_m$.

Let F be a field and let $\{\mathrm{X}_1, \mathrm{X}_2, \ldots, \mathrm{X}_n, \ldots\}$ be a set of countable infinite indeterminates over F. For each set $A_{m,k} \in \mathcal{P}_m$, let $\mathrm{Y}_{m,k} := \prod_{i \in A_{m,k}} \mathrm{X}_i$. For ease of notation, we let $\mathrm{Y}_{0,1} = \prod_{i \in \mathbb{N}} \mathrm{X}_i =: \mathrm{Y}$. Let $D_m := \bigcap_{k=1}^{m_n} V_{m,k}$ where $V_{m,k} := F[\mathrm{Y}_{m,1}, \mathrm{Y}_{m,2}, \ldots, \mathrm{Y}_{m,n_m}]_{(\mathrm{Y}_{m,k})}$. Set $R := \bigcup_{m=0}^{\infty} D_m$. From the construction it is obvious that $D_0 \subset D_1 \subset D_2 \cdots$ is an ascending chain of semilocal Dedekind domains. Moreover, each maximal ideal of D_m contracts to a maximal ideal of D_{m-1}. In particular, each contracts to the maximal ideal $\mathrm{Y}F[\mathrm{Y}]_{(\mathrm{Y})}$ in $D_0 = F[\mathrm{Y}]_{(\mathrm{Y})}$. We say that a family of sets $\mathcal{A} := \{A_{m,k_m}\}_{m=0}^{\infty}$ is a *chain through the series of partitions* $\mathcal{P} := \{\mathcal{P}_m\}_{m=0}^{\infty}$ if for each m, $A_{m,k_m} \supseteq A_{m+1,k_{m+1}}$. Depending on the choice of refinements $\mathcal{P}_m$, there may be chains through $\mathcal{P}$ which are eventually constant. As we will see, such a chain corresponds to a sharp prime of R.

Theorem 3.4.12. *Let R be as in Notation 3.4.11.*

(1) *If P is a nonzero maximal ideal of R, then $P \bigcap D_0 = \mathrm{Y}D_0$. Moreover, $PR_P = \mathrm{Y}R_P$.*
(2) *R is an almost Dedekind domain with nonzero Jacobson radical.*
(3) *Each finitely generated ideal of R is principal.*
(4) *There is a natural one-to-one correspondence between the set of maximal ideals of R and the set of chains through the family of partitions $\mathscr{P}$. Moreover, if M is a maximal ideal of M, then the corresponding chain of sets $\mathscr{A}$ is such that $\mathrm{Y}_{m,k_m} R_M = MR_M$ for each A_{m,k_m} in $\mathscr{A}$.*
(5) *The set $\{\mathrm{Y}_{m,k} \mid 0 \leq m,\ 1 \leq k \leq m_k\}$ contains the base set for a factoring family for R. Moreover, the set can be selected is such a way that each nonzero finitely generated fractional ideal will factor uniquely.*
(6) *A maximal ideal M of R is sharp if and only if the corresponding chain of sets $\mathscr{A}$ in statement (4) stabilizes at some $A_{m,k}$.*

Proof. Statements (1), (2) and (3) follow from Theorem 3.4.10. In particular, (3) is a result of Theorem 3.4.10(3) and the fact that each D_i is a PID. Statement (4) follows from the proof of Theorem 3.4.10(4) and the fact that each $\mathrm{Y}_{m,k}$ generates a maximal ideal of D_m. The statement in (5) follows from the proof of Theorem 3.4.10(5). Since each member of the factoring family is principal, each finitely generated ideal of R must be principal. Statement (6) is simply a combination of statement (4) and Theorem 3.4.10(4). □

This construction can be used to form almost Dedekind domains with various sharp degrees. Note that the domain R will have finite sharp degree if and only if there is an integer n such that R_n is semilocal.

We first show how to construct an almost Dedekind domain of sharp degree 2. The domain has a single dull maximal ideal.

Example 3.4.13. Let $\mathscr{P}_0 := \{\mathbb{N}\}$ and set $\mathscr{P}_m := \{A_{m,1}, A_{m,2}, \ldots, A_{m,m}, A_{m,m+1}\}$ for each $m \geq 1$ where $A_{m,i} = \{i\}$ for $1 \leq i \leq m$ and $A_{m,m+1} = \{k \in \mathbb{N} \mid k > m\}$. Let R be an almost Dedekind domain determined as in Notation 3.4.11 (and Theorem 3.4.12) by the series of partitions $\mathscr{P} := \{\mathscr{P}_m\}_{m=0}^{\infty}$ of $\mathbb{N}$. Then the following hold.

(1) R has exactly one maximal ideal M which is not sharp.
(2) R has sharp degree 2.
(3) R is a Bézout domain.
(4) $\mathrm{Max}^{\#}(R) = \{\mathrm{X}_n R \mid n \geq 1\}$ and the set $\{\mathrm{X}_n R \mid n \geq 1\} \bigcup \{\mathrm{Y}R\}$ is a factoring set for R such that each finitely generated ideal factors uniquely.
(5) There is a factoring family for R such that no nonzero finitely generated fractional ideal has a unique factorization over the underlying set of ideals.

Proof. (1)–(4) Let $\mathrm{Y}_n := \prod_{k=n+1}^{\infty} \mathrm{X}_k$ and hence, $\mathrm{Y}_0 = \prod_{k \in \mathbb{N}} \mathrm{X}_k =: \mathrm{Y}$. The maximal ideals of D_n consist of the ideal $\mathrm{Y}_n D_n$ and the ideals of the form $\mathrm{X}_k D_n$ for $1 \leq k \leq n$. (Note that, for $n = 0$, $D_0 = F[\mathrm{Y}]_{(\mathrm{Y})}$ is a local domain with maximal ideal (Y).) Thus for each integer $n \geq 1$, $\mathrm{X}_n R$ is a maximal ideal of R. Obviously, each of these is a sharp prime of R. The only other maximal ideal of R corresponds to the chain $\{\mathrm{Y}_n D_n\}$. Thus $R_2 = R_M$ where M is the maximal ideal of R determined by the chain $\{\mathrm{Y}_n D_n\}$ (corresponding to the chain $\mathscr{A} := \{\mathbb{N}, \{A_{m,m+1} \mid m > 0\}\}$ through $\mathscr{P}$).

Since M is the only dull prime of R and $\mathrm{Y}R_M = MR_M$ we have $\mathrm{Y}R_2 = MR_2$. By Theorem 3.4.9, the set $\{\mathrm{Y}R\} \bigcup \{\mathrm{X}_n R \mid n \geq 1\}$ is a factoring set for R such that each finitely generated fractional ideal factors uniquely over this set.

(5) For each $n \geq 1$, let $P_n := \mathrm{X}_n R$ and write $n = 4k - i$ where $k \geq 1$ and $0 \leq i \leq 3$. Build a factoring family for R as follows:

(a) for M again use $J_0 := \mathrm{Y}R$,
(b) if $i = 0$ (i.e., $n = 4k$), let $J_n := \mathrm{X}_{2k-1}^3 \mathrm{Y}R$,
(c) if $i = 1$ (i.e., $n = 4k - 1$), let $J_n := \mathrm{X}_{2k}^3 \mathrm{Y}R$,
(d) if $i = 2$ (i.e., $n = 4k - 2$), let $J_n := \mathrm{X}_{2k-1}^2 \mathrm{Y}R$, and
(e) if $i = 3$ (i.e., $n = 4k - 3$), let $J_n := \mathrm{X}_{2k}^2 \mathrm{Y}R$.

Note that, obviously, $\mathrm{X}_m R$ is the product of $(\mathrm{X}_m^3 \mathrm{Y}R)(\mathrm{X}_m^2 \mathrm{Y}R)^{-1}$. Hence the set $\{J_n\}_{n=0}^{\infty}$ is a factoring family for R. But factorizations are not unique. For example, $\mathrm{X}_m R$ can also be factored as $(\mathrm{X}_m^2 \mathrm{Y}R)^2 (\mathrm{X}_m^3 \mathrm{Y}R)^{-1} (\mathrm{Y}R)^{-1}$. There are in fact infinitely many different ways to factor each nonzero finitely generated fractional ideal of R. By the construction of the family, it is clear that each factorization of $\mathrm{X}_m R$ must contain nonzero powers of both $\mathrm{X}_m^2 \mathrm{Y}R$ and $\mathrm{X}_m^3 \mathrm{Y}R$. On the other hand, $\mathrm{Y}R$ is redundant, as it can be factored as $(\mathrm{X}_m^2 \mathrm{Y}R)^2 (\mathrm{X}_m^3 \mathrm{Y}R)^{-1}$. □

Next, we construct an almost Dedekind domain for which each maximal ideal is dull and where at least some finitely generated ideals will fail to factor uniquely over whatever factoring family we might use, but not necessarily fail to factor uniquely over the underlying set of potential factors.

Example 3.4.14. Let $\mathscr{P}_0 := \{\mathbb{N} =: A_{0,1}\}$ and let $\mathscr{P}_n := \{A_{n,1}, A_{n,2}, \cdots, A_{n,2^n}\}$ for each $n \geq 1$ where $A_{n,k} := \{m2^n + k \mid m \in \mathbb{Z}, m \geq 0\}$ for each integer $1 \leq k \leq 2^n$. Let R be the almost Dedekind domain determined as in Notation 3.4.11 (and Theorem 3.4.12).

(1) R is an almost Dedekind domain which is dull.
(2) There exists a factoring family $\{J_\alpha\}$ such that each nonzero finitely generated ideal factors uniquely over the underlying set of ideals making up the family.
(3) Given any factoring family $\{J_\alpha\}$ for R, there exists a nonzero finitely generated ideal I which does not factor uniquely over the family.

Proof. (1) As no chain of sets through $\mathscr{P}$ stabilizes, R has no sharp primes. Hence R is an almost Dedekind dull domain.

(2) By the proof of Theorem 3.4.10(5) (or Theorem 3.4.12(5)), some subset of $\{\mathrm{Y}_{m,k}\}$ contains a set such that (i) each nonzero finitely generated fractional ideal factors uniquely, and (ii) this set is the underlying set for a factoring family for R.

(3) The nonuniqueness is simply a consequence of the fact that R has only countably many nonzero finitely generated fractional ideals, but an uncountable number of maximal ideals. Thus for each factoring family, at least two members are the same ideal of R.

□

It is actually rather easy to modify the construction in Example 3.4.14 to obtain an almost Dedekind domain R of dull degree two. One quite trivial way is to simply replace each set $A_{r,1}$, with $r \geq 1$, by the sets $\{1\}$ and $\{m2^r + 1 \mid m \in \mathbb{N}\}$. This will yield exactly one sharp prime, with the rest dull, and therefore destined to stay that way in R_2. For a more elaborate example with infinitely many sharp primes, we modify the $\mathscr{P}_r$s a bit more.

Example 3.4.15. Start with the partitions $\mathscr{P}_n$ of Example 3.4.14. Then, for each n and each $0 \leq r \leq n$, split each set $A_{n,2^r}$ into the singleton set $\{2^r\}$ and the set $A'_{n,2^r} := \{m2^n + 2^r \mid m \in \mathbb{N}\}$. The new $\mathscr{P}_n$ consists of the singleton sets $\{2^r\}$ for $0 \leq r \leq n$, the sets $A'_{n,2^r}$ and the previous sets $A_{n,k}$ when $k < 2^n$ is not a power of 2. Let R be almost Dedekind domain determined as in Notation 3.4.11 (and Theorem 3.4.12) by the chains through the series of partitions $\mathscr{P} := \{\mathscr{P}_n\}_{n=0}^{\infty}$ of $\mathbb{N}$. Then R is an almost Dedekind domain with infinitely many sharp primes and dull degree 2.

Proof. Obviously, each singleton set $\{2^r\}$ corresponds to a sharp prime $M_r R = X_{2^r} R$. Each of these primes blows up in $R_2 (= \bigcap\{R_N \mid N \in \mathrm{Max}^{\dagger}(R)\})$, the effect is the same as beginning the construction by partitioning the set $\mathbb{N}\backslash\{2^r \mid r \geq 0\}$ as in Example 3.4.14. Thus R_2 is a dull domain. □

To construct almost Dedekind domains of larger sharp and dull degrees, we essentially take a recursive approach. The basic idea is to shift the partitioning scheme used to produce a domain with sharp/dull degree n in such a way as to increase the sharp/dull degree up to $n+$. To make this precise we introduce some useful terminology. Given a set $A_{m,k}$ in a chain of partitions, we consider the family of sets $\{A_{n,j} \mid A_{n,j} \subseteq A_{m,k}, n \geq m\}$ and call this the *branch of the partition from* $A_{m,k}$. Such a branch is said to have *sharp degree p*, if each maximal ideal which has $A_{m,k}$ in its corresponding family of sets has sharp degree less than or equal to p and at least one such maximal ideal has sharp degree p. On the other hand, a branch is said to have *dull degree p*, if there is a maximal ideal which has $A_{m,k}$ in its corresponding family of sets that is dull in every R_n, but there are maximal ideals of sharp degree $p - 1$ corresponding to the same $A_{m,k}$, but none of higher sharp degree.

To build a branch of sharp degree two we may use a scheme quite similar to that used in Example 3.4.13. Let $\{\mathscr{P}_m\}$ be a series of refinements. For ease of notation assume that for each pair of integers $m < n$, the set $A_{m,1}$ is infinite and $A_{m,1}$ contains $A_{n,1}$. Fix m and order the elements of $A_{m,1}$ as $a_1 < a_2 < a_3 < \ldots$. Then, as in Example 3.4.13, for each integer $n > m$, let $A'_{n,1} := \{a_1\}, A'_{n,2} := \{a_2\}, \ldots, A'_{n,n-m} := \{a_{n-m}\}$ and let $A'_{n,n-m+1}$ be the rest of $A_{m,1}$. In each $\mathscr{P}_n$, replace the sets which contain $A_{m,1}$ by the $A'_{n,j}$ sets and leave the rest of $\mathcal{P}_n$ as it is. Then there is exactly one maximal ideal M of (the new) R, the almost Dedekind domain determined as in Notation 3.4.11 (and Theorem 3.4.12), whose corresponding chain contains $A_{m,1}$ and is not sharp, the one associated with the sets $A'_{n,n-m+1}$. All other maximal ideals associated with $A_{m,1}$ have chains which stabilize at some singleton set $\{a_r\}$. We refer to this technique as building a *standard branch of sharp degree two*.

In our next example we utilize this basic construction to build an almost Dedekind domain of sharp degree 3. The construction of the partitions is more complicated, so we will give the details of the construction in the proof rather than the statement of what we are going to build.

Example 3.4.16. There is a series of partitions $\mathcal{P} = \{\mathcal{P}_m\}_{m=0}^{\infty}$ such that the resulting domain R, determined as in Notation 3.4.11 (and Theorem 3.4.12), is an almost Dedekind domain having a unique maximal ideal M with sharp degree 3, so $R_3 = R_M$ and R has sharp degree 3.

Let $\mathcal{P}_0 := \{\mathbb{N}\}$ and $\mathcal{P}_1 := \{E, O\}$, where E denotes the positive even integers and O denotes the positive odd integers. From O, build the standard branch of sharp degree two. But for E we proceed a little differently. First split E into the sets $E_{4,0} := \{4m \mid m \geq 1\}$ and $E_{4,2} := \{4m+2 \mid m \geq 0\}$. From $E_{4,2}$ build the standard branch of sharp degree two, but split $E_{4,0}$ into sets $E_{8,0} := \{8m \mid m \geq 1\}$ and $E_{8,4} := \{8m+4 \mid m \geq 0\}$. Then, as with $E_{4,2}$, build the standard branch of sharp degree two from $E_{8,4}$, and, as with $E_{4,0}$, split $E_{8,0}$ into sets $E_{16,0} := \{16m \mid m \geq 1\}$ and $E_{16,8} := \{16m+8 \mid m \geq 0\}$. Continue this scheme for each power of 2. Let R be the resulting almost Dedekind domain and let M be the maximal ideal corresponding to the chain $\{E_{2^n,0}\}$.

Proof. We will show that there is one prime of sharp degree two associated with O and that each set $E_{2^n,2^{n-1}}$ is associated to exactly one prime of sharp degree two.

The only sharp primes of R are those associated with some singleton set $\{a\}$. For each positive integer n, there is exactly one prime of sharp degree two that contains $\prod_{r=0}^{\infty} X_{2^n r+2^{n-1}}$, the one associated with the chain $\{B^{m,n}\}_{m=1}^{\infty}$ where $B^{m,n} := \{2^n r + 2^{n-1} \mid r \geq m\}$. On the other hand the chain associated with M consists of the sets of the form $\{2^n r \mid n \geq 0, r \geq 1\}$, so $\mathbb{N}$, E, $E_{4,0}$, $E_{8,0}$, etc. For each n, there are infinitely many primes of sharp degree two which are associated with $E_{2^n,0}$. Hence M cannot have sharp degree two. As it is the only dull prime which does not have sharp degree two, it must have sharp degree three. Thus R has sharp degree three and $R_3 = R_M$. □

Theorem 3.4.17. *For each positive integer $k \geq 2$, there is a series of refinements $\{\mathcal{P}_m\}$ of $\mathcal{P}_0 := \{\mathbb{N}\}$ such that the resulting domain R, determined as in Notation 3.4.11 (and Theorem 3.4.12), is an almost Dedekind domain of sharp degree k.*

Proof. The proof is by induction on $k \geq 2$. Assume the result holds for k. The partitioning scheme is somewhat a combination of those used in Examples 3.4.14 and 3.4.16. As in Example 3.4.14, we let $\mathcal{P}_1 := \{O, E\}$ and $\mathcal{P}_2 := \{A_{2,1}, A_{2,2}, A_{2,3}, A_{2,4}\}$ with each $A_{2,r} := \{m2^2 + r \mid m \geq 0\}$. The subsequent partitions will be different. Specifically, from $A_{2,2}$ and $A_{2,3}$ build branches of sharp degree k. On the other hand, we split $A_{2,1}$ into $A_{3,1}$ and $A_{3,5}$ and split $A_{2,4}$ into $A_{3,4}$ and $A_{3,8}$ as in the third stage of the process in Example 3.4.14. Now continue the pattern of splitting the sets $A_{n,2^n}$ and $A_{n,1}$ as in Example 3.4.14, but split the sets $A_{n,2^{n-1}}$ and $A_{n,2^{n-1}+1}$ into branches of sharp degree k. Each branch of the infinitely many branches of sharp degree k corresponds to maximal ideals of sharp degree k. But, the prime associated with the chain $\{A_{n,2^n}\}$ will not have sharp degree k, since each of the sets $A_{n,2^n}$ is in infinitely many chains associated with primes of sharp degree k. The same is true for the prime associated with the chain $\{A_{n,1}\}$. As these are the only chains which do not lead to primes of sharp degree less than or equal to k, each has sharp degree $k+1$ and therefore R is an almost Dedekind domain of sharp degree $k+1$. □

We take a slightly different approach in increasing dull degree. Instead of splitting sets into two nonempty subsets, we split them into three. Also, we allow infinite sets to stabilize. We start with an example illustrating how to use thirds to build an almost Dedekind domain of dull degree two with infinitely many sharp primes. The basic construction parallels the "excluded middle" construction of a Cantor set. This makes it rather easy to increase the dull degree. Our first task is to create an almost Dedekind domain of dull degree 2 that has infinitely many invertible (= sharp) maximal ideals.

Example 3.4.18. We make use of trinary expansions of integers. For each pair of integers $n \geq 1$ and $1 \leq r \leq 3^n$, we set $A_{n,r} := \{m3^n + r \mid m \geq 0\}$ and let $r =: r_n r_{n-1} \ldots r_1$ be the trinary expansion of r. We start with $\mathscr{P}_0 := \{\mathbb{N}\}$ and then for $n \geq 1$ we let $\mathscr{P}_n := \{A_{n,r} \mid \text{no } r_i \text{ is a } 2\} \bigcup \{A_{k,s} \mid 1 \leq k \leq n \text{ is the smallest integer such that } s_k = 2\}$. The resulting domain R, determined as in Notation 3.4.11 (and Theorem 3.4.12), has dull degree two with infinitely many sharp primes.

Proof. We start with an explicit construction for the first few $\mathscr{P}_n$s. First $\mathscr{P}_1 = \{A_{1,1}, A_{1,2}, A_{1,3}\}$. Then, for $\mathscr{P}_2$, we leave the set $A_{1,2}$ as is but split $A_{1,1}$ into $A_{2,1}$, $A_{2,4}$ and $A_{2,7}$, and split $A_{1,3}$ into $A_{2,3}$, $A_{2,6}$ and $A_{2,9}$. The set $A_{1,2}$ will appear in each $\mathscr{P}_n$ from here on, as will the sets $A_{2,4}$ and $A_{2,6}$. On the other hand, we split $A_{2,1}$ into $A_{3,1}$, $A_{3,10}$, and $A_{3,19}$, $A_{2,3}$ into $A_{3,3}$, $A_{3,12}$ and $A_{3,21}$, $A_{2,4}$ into $A_{3,4}$, $A_{3,13}$ and $A_{3,22}$, and $A_{2,9}$ into $A_{3,9}$, $A_{3,18}$ and $A_{3,27}$. In $\mathscr{P}_4$, we simply keep each "middle third" as it is and split each pair of outer thirds based on the remainders on division by 3^4. Continue this process to build the partitions $\mathscr{P}_n$. As each middle third set is stable once it appears in some $\mathscr{P}_n$, each leads to a sharp prime of R. On the other hand, if the chain of sets corresponding with a maximal ideal M of R contains no middle third set, then each set in the chain is associated with many infinitely many maximal ideals, including infinitely many which are not associated with a middle third set. Thus R has dull degree 2 with infinitely many sharp primes. □

In the proof for the next theorem, we show how the construction in the previous example can be used to construct an almost Dedekind domain of arbitrary (finite) dull degree $k \geq 2$.

Theorem 3.4.19. *For each integer $k \geq 1$, there exists an almost Dedekind domain of dull degree k.*

Proof. Examples 3.4.14 and 3.4.15 provide almost Dedekind domains of dull degree one and two, respectively. As in Theorem 3.4.17, we modify a previous construction by taking out sets which have stabilized and replacing them with branches of the appropriate sharp degree. Our construction is based on that in Example 3.4.15.

Fix $k \geq 3$. The outer third sets are left as they are in Example 3.4.15, but each middle third set is replaced by a branch of sharp degree $k - 1$. Each of the new chains will lead to a maximal ideal of sharp degree $k - 1$ or less, with infinitely many of sharp degree $k - 1$. This is the maximal sharp degree of any maximal ideal of R, determined as in Notation 3.4.11 (and Theorem 3.4.12). Each prime resulting from a chain of outer third sets remains dull in R_k. Thus R_k is a dull domain, with R_{k-1} a proper subring. Hence R has dull degree k. □

Theorem 3.4.20. *There exists an almost Dedekind domain R such that R_n is a proper subring of R_{n+1} for each integer positive integer n. Moreover, the ring $R_\infty := \bigcup R_n$ may be a sharp domain, a dull domain or have some other sharp or dull degree.*

Proof. We start with constructing a domain R such that R_∞ has sharp degree one with $R_n \neq R_{n+1}$ for each n. For this purpose, start with the basic Odd/Even partitioning scheme used to construct branches of sharp degree k, but instead of changing each branch to one of sharp degree $k - 1$, allow each new branch to have larger and larger sharp degree. By doing so, once we hit a set high enough up in the branch of sharp degree n, we find a single prime of sharp degree n and all others with smaller sharp degree. But now, the chain corresponding to the powers of 2 sets will not lead to a prime of finite sharp degree. However, once we take the union of the R_ns, we will obtain a domain of sharp degree one as the only prime which does not have finite sharp degree is the one corresponding to the chain $\{E_{n,2^n}\}$.

We use a similar scheme to build a domain R such that R_∞ is a dull domain with primes of each finite sharp degree. Start with the basic scheme used in the proof of Theorem 3.4.17, but now instead of replacing each middle third set with a branch of the same sharp degree, replace them with branches of larger and larger sharp degree. We may leave the first middle third set, $A_{1,2}$, alone. Then replace $A_{2,4}$ and $A_{2,6}$ by branches of sharp degree two. Continue by replacing each middle third set $A_{k,r}$ by a branch of sharp degree k. The result will be that each branch through a middle third set leads only to primes of finite sharp degree, but there is no uniform bound on the degree that holds for all branches through all middle third sets. As in the proof of Theorem 3.4.17, the primes whose chains involve only outer third sets will remain dull throughout each R_n and remain dull in R_∞. Thus R_∞ is a dull domain.

For sharp and dull degree two for R_∞, replace branches of finite sharp degree with ones which mimic the construction of a R_∞ with sharp degree one. Continue this fractal like approach to get larger and larger sharp and dull degrees for R_∞. □

Chapter 4
Weak, Strong and Very Strong Factorization

Abstract An integral domain is said to have weak factorization if each nonzero nondivisorial ideal can be factored as the product of its divisorial closure and a finite product of (not necessarily distinct) maximal ideals. An integral domain is said to have strong factorization if it has weak factorization and the maximal ideals of the factorization are distinct. If, in addition, the maximal ideals in the factorization of a nonzero nondivisorial ideal I of the domain R can be restricted to those maximal ideals M such that IR_M is not divisorial, we say that R has very strong factorization. In the present section, we study these properties with particular regard to the case of Prüfer domains or almost Dedekind domains. In the Prüfer case we provide several characterizations of domains having weak, strong or very strong factorization. We discuss the connections with h-local domains and we prove that very strong and strong factorizations are equivalent for Prüfer domains.

4.1 History

In [19], the authors introduced two factorization properties for integral domains. We start by recalling the first one, called "weak" factorization.

An integral domain R is said to have *weak factorization* if each nonzero nondivisorial ideal I can be factored as the product of its divisorial closure I^v and a finite product of (not necessarily distinct) maximal ideals; i.e.,

$$I = I^v M_1 M_2 \cdots M_n\,, \quad \text{where } M_i \in \operatorname{Max}(R) \text{ for } 1 \leq i \leq n.$$

In [19], the second factorization was called "strong" factorization and had two additional restrictions; first, the maximal ideals $\{M_1, M_2, \ldots, M_n\}$ in the factorization of a nonzero nondivisorial ideal I were required to be distinct, and, second, for $I = I^v M_1 M_2 \cdots M_n$, the M_i had to be precisely those maximal ideals M for which IR_M is not a divisorial ideal of R_M.

M. Fontana et al., *Factoring Ideals in Integral Domains*, Lecture Notes of the Unione Matematica Italiana 14, DOI 10.1007/978-3-642-31712-5_4,

It turns out that for Prüfer domains, there is no need to include this second requirement (see Theorem 4.4.8) below. Thus we now redefine two types of "strong" factorization by distinguishing, a priori, two possible situations. It is convenient to let $\mathscr{H}(I)$ denote the (possibly empty) set of maximal ideals M such that $IR_M \neq (IR_M)^v$. (So formally, $\mathscr{H}(I) = \{M \in \text{Max}(R) \mid IR_M \neq I^v R_M\}$.)

An integral domain R is said to have *strong factorization* if each nonzero nondivisorial ideal I of R can be factored as follows:

$$I = I^v M_1 M_2 \cdots M_n\,, \quad \text{where } M_i \in \text{Max}(R) \text{ and } M_i \neq M_j \text{ for } 1 \leq i \neq j \leq n.$$

If, in addition, the M_i in such a factorization can be restricted to those maximal ideals M such that IR_M is not divisorial, we say that R has *very strong factorization*. That is, R has very strong factorization if for each nonzero nondivisorial ideal I, I can be factored as follows:

$$I = I^v M_1 M_2 \cdots M_n, \text{ where } \mathscr{H}(I) = \{M_1, M_2, \ldots, M_n\} (\neq \emptyset).$$

Remark 4.1.1. It is rather trivial to show (by checking locally) that in any of these factorizations, if $IR_{M_i} = I^v R_{M_i}$ for some M_i, then it must be that $I^v M_i = I^v$ and thus the factor of M_i can be eliminated.

One of the main theorems of [19] is the following.

Theorem 4.1.2. [19, Theorem 1.12] *The following statements are equivalent for a Prüfer domain R.*

(i) R is h-local.
(ii) R has the very strong factorization property.
(iii) For each nonzero ideal I of R, I is divisorial if and only if IR_M is divisorial in R_M for each maximal ideal M of R.
(iv) For each nonzero ideal I of R, if IR_M is divisorial for each maximal ideal M, then I is divisorial.

In Sect. 4.4, we prove a sharper version of the equivalence (i)⇔(ii) of Theorem 4.1.2. More precisely, in Theorem 4.4.9, we show that if R is an integral domain that possesses (the new type of) strong factorization, then each nonzero finitely generated ideal is divisorial. As a corollary, we have that if R is integrally closed, then it has our redefined form of strong factorization if and only if it is an h-local Prüfer domain (Corollary 4.4.10); i.e., in this situation, very strong factorization and strong factorization coincide.

One of the key results used in the proof of the Theorem 4.1.2 is the following Proposition 4.1.3 [19, Theorem 1.10]. Its expanded version, Theorem 2.5.2, plays a significant role in proving several of the results to come.

Proposition 4.1.3. *Let R be a Prüfer domain and let P be a nonzero nonmaximal prime that is the radical of a finitely generated ideal. If I is a finitely generated ideal whose radical is P and M is a maximal ideal that contains P, then the* *ideal* $J := IR_M \bigcap R$ *is divisorial if and only if* M *is the only maximal ideal that contains* P.

As a consequence of Theorems 2.2.1, 2.4.12 and 4.1.2, we have:

Corollary 4.1.4. *An almost Dedekind domain with very strong factorization is Dedekind.*

However, there exist almost Dedekind domains with weak factorization that do not have (very) strong factorization. More precisely, we proved in [19, Theorem 1.15] the following.

Theorem 4.1.5. *Let* R *be an almost Dedekind domain, and let* I *be a nonzero ideal of* R *which is contained in only finitely many nondivisorial maximal ideals of* R. *Then* $I = I^v M_1 M_2 \cdots M_n$ *where the* M_i *are maximal ideals but are not necessarily distinct. Thus if* R *is an almost Dedekind domain in which each nonzero ideal is contained in only finitely many nondivisorial maximal ideals, then* R *has the weak factorization property.*

By using Theorem 4.1.5, [34, Example 42.6] provides an explicit example of an almost Dedekind domain with weak factorization that does not have very strong factorization. Other examples can be found in [58].

Below, in Proposition 4.2.14, we will give several ways of characterizing when an almost Dedekind domain that is not Dedekind has weak factorization, essentially establishing the converse of Corollary 4.1.4.

4.2 Weak Factorization

In Theorem 4.1.2, Prüfer domains which are h-local were characterized via the very strong factorization property. On the other hand, in the Prüfer domain case, h-local domains can be also characterized using the weak factorization property. More precisely, [19, Theorem 1.13] provides the following characterization.

Theorem 4.2.1. *Let* R *be a Prüfer domain. Then* R *is* h*-local if and only if* R *has weak factorization and finite character.*

In Theorem 4.4.8 below, we will give another proof of this, together with several other new characterizations of h-local Prüfer domains based on weak factorization-type properties.

By [19, Proposition 1.7], a Prüfer domain with weak factorization is a wTPP-domain (Sect. 2.4) and, more precisely, we have the following.

Proposition 4.2.2. *Let* R *be a Prüfer domain with the weak factorization property. Then the following hold.*

(1) each ideal which is primary to a nonmaximal ideal of R *is divisorial (in particular, each nonmaximal prime is divisorial),*

(2) *if M is an idempotent maximal ideal of R and I is a nondivisorial M-primary ideal, then $I = I^v M$,*
(3) *each branched maximal idempotent ideal of R is sharp,*
(4) *R is a wTPP-domain, and*
(5) *each branched nonmaximal prime ideal of R is the radical of a finitely generated ideal.*

The next lemma collects a few useful properties a Prüfer domain with weak factorization property has in common with one having very strong factorization (for comparison, see [67, Proposition 3.4 and Theorem 3.10] and [19, Proposition 2.10]).

Lemma 4.2.3. *Let R be a Prüfer domain with the weak factorization property.*

(1) *Each nonzero prime is contained in a unique maximal ideal.*
(2) *Each maximal ideal of height greater than one is sharp.*
(3) *Each locally principal maximal ideal of height greater than one is invertible. Thus (equivalently) each unsteady maximal ideal has height one.*
(4) *If I is a nonzero, nondivisorial ideal with factorization $I = I^v \prod N_i \prod M_j^{r_j}$ where the N_i are the steady maximal ideals for which $IR_{N_i} \neq I^v R_{N_i}$ and the M_j are the unsteady maximal ideals for which $IR_{M_j} \neq I^v R_{M_j}$, then each N_i is idempotent and $I^v R_{N_i}$ is principal.*

Proof. As each unbranched prime contains a nonzero branched prime, it suffices to prove (1) in the case $P \neq (0)$ is a nonmaximal branched prime. By Proposition 2.3.10, there is a finitely generated ideal I such that $\sqrt{I} = P$. Let M be a maximal ideal that contains P and let $J := IR_M \bigcap R$. Then by Theorem 2.5.2(2) (enlarging I if necessary), $J^v = J(P' : P') = I(P' : P')$ where P' is the largest prime that is common to all maximal ideals that contain P. Clearly, $\text{Max}(R, P) = \text{Max}(R, P')$. Thus P' is sharp by Lemma 2.3.9. Therefore P' is a maximal ideal of $(P' : P')$ by Corollary 2.3.21(2). Moreover, P' is the only maximal ideal of $(P' : P')$ that contains P, I and J^v (since such a maximal ideal must be extended from a prime P'' of R which must be contained in one of the maximal ideals containing I and must therefore be comparable to P'). Hence $J^v R_{P'} = IR_{P'} = JR_{P'}$ and $J^v = JR_{P'} \bigcap (P' : P')$. (The latter equality is true locally since P' is the only maximal ideal of $(P' : P')$ which contains J and J^v.) It follows that $J^v = JR_{P'} \bigcap R$.

Since R has weak factorization and $JR_{P'} = J^v R_{P'}$, there is an ideal H that is not contained in P' such that $J = J^v H$ (either with $H = R$ or H a finite product of maximal ideals). By Lemma 2.5.1(2), $JR_{P'} = J^v \Gamma(P)$ and this yields $IR_M = JR_M \supseteq J\Gamma(P) = J^v H\Gamma(P) = JHR_{P'} = JR_{P'} = IR_{P'} \supseteq IR_M$. Hence $IR_M = IR_{P'}$. Since I is finitely generated, we must have $M = P'$. This establishes (1).

For (2), (3) and (4), let M be a maximal ideal of height greater than one. Then M contains at least one nonzero branched prime. For (2), simply apply (1), Propositions 2.3.10 and 4.2.2(5), and Lemma 2.3.9 to see that M is sharp. For (3), further assume that M is locally principal, say $MR_M = aR_M$ with $a \in M$.

Since M is sharp, there is a finitely generated ideal A of R such that M is the only maximal ideal that contains A. Then the ideal $A + aR$ generates M locally, whence $M = A + aR$ is invertible.

Finally, (4) follows from Lemma 2.5.3. □

As noted earlier in Sect. 2.4, a Prüfer domain with weak factorization is an aRTP-domain. Using Lemma 4.2.3 and several of the other results above, we are now ready to give our first alternate characterizations of Prüfer domains with weak factorization.

Theorem 4.2.4. *The following statements are equivalent for a Prüfer domain R.*

(i) R has weak factorization.

(ii) Each nonzero prime is contained in a unique maximal ideal, and R is an aRTP-domain such that for each nonzero ideal I, the set of maximal ideals N where $IR_N \neq I^v R_N$ is finite.

(iii) (a) Each steady maximal ideal is sharp,

(b) each nonzero nonmaximal prime ideal is sharp and contained in a unique maximal ideal, and

(c) for each nonzero ideal I, the set of maximal ideals N where $IR_N \neq I^v R_N$ is finite.

Proof. By Theorem 2.4.18, a Prüfer domain is an aRTP-domain if and only if each nonzero branched nonmaximal prime ideal and each steady branched maximal ideal are sharp. Under the additional assumption that each nonzero prime is contained in a unique maximal ideal, Lemma 2.3.9 guarantees that the equivalence holds true with the word "branched" removed. Hence (ii) and (iii) are equivalent.

To see that (i) implies (ii), assume that R has weak factorization. If I is a nonzero nondivisorial ideal, then we have $I = I^v \prod_{i=1}^n M_i^{s_i}$ for some finite set of maximal ideals $\{M_1, M_2, \ldots, M_n\}$ and positive integers $s_1, s_2, \ldots, s_n$. From this it is clear that there are at most finitely many maximal ideals N such that $IR_N \neq I^v R_N$. Also, by Lemma 4.2.3(1), each nonzero prime is contained in a unique maximal ideal. Finally, each branched idempotent maximal ideal and each nonzero nonmaximal branched prime ideal are sharp by Proposition 4.2.2(3 and 5), so that R is an aRTP domain by Theorem 2.4.18.

To complete the proof we show (iii) implies (i). Assume all three conditions in (iii) hold. By Lemma 2.3.9, statement (iii)(b) implies that each maximal ideal of height greater than one is sharp. Combined with (iii)(a), we have that the only maximal ideals that are not sharp are the height one unsteady maximal ideals.

Let I be a nonzero nondivisorial ideal. Then by (iii)(c), there is a nonempty finite set of maximal ideals $\{M_1, M_2, \ldots, M_n\}$ such that $IR_{M_i} \neq I^v R_{M_i}$ for $1 \leq i \leq n$ and $IR_M = I^v R_M$ for all maximal ideals M not in the set $\{M_1, M_2, \ldots, M_n\}$. If M_i is a height one unsteady maximal ideal, then $M_i R_{M_i}$ is principal and therefore $IR_{M_i} = M_i^{r_i} R_{M_i} \subsetneq I^v R_{M_i} = M^{t_i} R_{M_i}$ for some integers $r_i > t_i \geq 0$. In this case, $IR_{M_i} = I^v M_i^{s_i} R_{M_i}$, where $s_i = r_i - t_i$. For those M_j that are not height one unsteady maximal ideals, Theorem 2.5.4 (together with statements (iii)(a) and (iii)(b)) implies that M_j must be idempotent with $IR_{M_j} = I^v M_j R_{M_j}$. By checking locally, we have $I = I^v \prod_{i=1}^n M_i^{s_i}$ for some positive integers $s_1, s_2, \ldots, s_n$. Hence (iii) implies (i). □

Theorem 4.2.5. *Let R be a Prüfer domain with the weak factorization property that is not h-local. If S is an overring of R where no unsteady maximal ideal of R survives, then S is h-local.*

Proof. By Theorems 2.4.18 and 4.2.4, if P is a nonzero branched prime ideal of R that is not an unsteady maximal ideal, then P is the radical of finitely generated ideal and it is contained in a unique maximal ideal.

Assume that S is an overring of R. Then each nonzero prime of S is contained in a unique maximal ideal. If no unsteady maximal ideal of R survives in S and Q is a (nonzero) branched prime of S, then Q must be extended from either a steady maximal ideal or a branched nonmaximal prime. In either case, $Q \bigcap R$ is the radical of a finitely generated ideal and therefore so is Q. Thus S has the radical trace property by Theorem 2.4.14. Therefore S is h-local by Theorem 2.4.12. □

Corollary 4.2.6. *Let R be a Prüfer domain with the weak factorization property. If I is an ideal that is contained in no unsteady maximal ideals, then I is contained in only finitely many maximal ideals.*

Proof. Suppose I is contained in no unsteady maximal ideals. Then each maximal ideal that contains I is sharp (Theorem 4.2.4). Hence by Lemmas 2.5.1 and 2.4.19, the ring $\Gamma(I)$ has no unsteady maximal ideals and each maximal ideal of $\Gamma(I)$ contains I. Since R has the weak factorization property, $\Gamma(I)$ is h-local by Theorem 4.2.5, and so, in particular, $\Gamma(I)$ has finite character. Thus at most finitely many maximal ideals of R contain I. □

Theorem 4.2.7. *Let R be a Prüfer domain with weak factorization. If I is a radical ideal of R such that I^{-1} is a ring, then each minimal prime of I extends to a maximal ideal of I^{-1} as does each maximal ideal of R that does not contain I.*

Proof. Assume that I is a radical ideal such that I^{-1} is a ring. Since $\Theta(I)$ contains I^{-1} (Theorem 2.3.2(1)), it is always the case that a maximal ideal that does not contain I extends to a maximal ideal of I^{-1}. Suppose that P is a prime minimal over I. If P is nonmaximal, then it is (ante)sharp by Theorem 4.2.4 and Corollary 2.3.21(2). If P is maximal, then it is trivially antesharp. Hence PI^{-1} is a maximal ideal of I^{-1} by Lemma 2.5.5(1). □

We are primarily interested in applying Theorem 2.5.6 in the case that R is a Prüfer domain with weak factorization. For this situation, we can make a slight change in the hypothesis.

Theorem 4.2.8. *Let R be a Prüfer domain with weak factorization, and let P be a sharp prime of R. If I is radical ideal with $I \subseteq P$ and $\{P_\alpha\}$ is a set of minimal prime ideals such that $I = \bigcap_\alpha P_\alpha$, then P contains some $P_\beta \in \{P_\alpha\}$. If, in addition, P is not minimal over I, then $P \bigcap (\bigcap_{\alpha \neq \beta} P_\alpha)$ properly contains I.*

Proof. Simply apply Theorem 2.5.6 to the prime $Q \subseteq P$ with Q minimal over I. Such a prime is sharp since R has weak factorization (Theorem 4.2.4). □

The next corollary collects several useful consequences of Theorems 2.5.6 and 4.2.8.

Corollary 4.2.9. *Let R be a Prüfer domain with weak factorization.*

(1) If $\mathscr{W} := \{M_\alpha \mid \alpha \in \mathscr{A}\}$ is a nonempty set of unsteady maximal ideals such that $I := \bigcap_\alpha M_\alpha$ is a nonzero ideal, then no sharp prime contains I.
(2) If $\{P_\alpha\}$ is a nonempty set of pairwise incomparable sharp primes such that $J := \bigcap_\alpha P_\alpha \neq (0)$, then each P_α is minimal over J and no other sharp prime is minimal over J.
(3) No nonzero element of R is contained in infinitely many idempotent maximal ideals.

Proof. Let $I = \bigcap_\alpha M_\alpha$ be nonzero, with $\mathscr{W} = \{M_\alpha \mid \alpha \in \mathscr{A}\}$ a nonempty set of unsteady maximal ideals. Then each $M_\alpha \in \mathscr{W}$ has height one (Lemma 4.2.3(3)) and is therefore minimal over I. Moreover, the M_α are not sharp by Lemma 2.4.19. Hence no sharp prime contains I by Theorem 4.2.8, proving (1).

For (2), assume that $\{P_\alpha\}$ is a nonempty set of pairwise incomparable sharp primes such that $J = \bigcap_\alpha P_\alpha \neq (0)$. For each α, let $Q_\alpha \subseteq P_\alpha$ be a prime ideal that is minimal over J. Obviously, $J = \bigcap_\alpha Q_\alpha$ and for each P_β, $J = \bigcap_\alpha Q_\alpha \subseteq P_\beta \bigcap (\bigcap_{\alpha \neq \beta} Q_\alpha) \subseteq P_\beta \bigcap (\bigcap_{\alpha \neq \beta} P_\alpha) = J$. Thus by Theorem 4.2.8, each P_α is minimal over J, and no other sharp prime is minimal over J.

To see that (3) holds, further assume that each P_α in the intersection $J = \bigcap_\alpha P_\alpha$ of (2) is an idempotent maximal ideal. (Note that each P_α is sharp by Theorem 4.2.4.) Since $JR_{P_\alpha} = P_\alpha R_{P_\alpha}$ and P_α is idempotent, it must be that $JJ^{-1}R_{P_\alpha} = P_\alpha R_{P_\alpha}$ (since $JR_{P_\alpha} \subseteq JJ^{-1}R_{P_\alpha} \subseteq P_\alpha R_{P_\alpha}(P_\alpha R_{P_\alpha})^{-1} = P_\alpha R_{P_\alpha} = JR_{P_\alpha}$). Thus $JJ^{-1} = J$, and we have $J^{-1} = (J : J)$, and so J^{-1} is a ring. By Theorem 4.2.8, no other sharp prime can be minimal over J, whence by Lemma 4.2.3, the only other minimal primes of J must be height one (unsteady) maximal ideals. Thus $J^{-1} = \Gamma(J) \bigcap \Theta(J) = R$ (Theorem 2.3.2(2)) and so $J \neq J^v = R$. It follows that the set $\{P_\alpha\}$ is finite by Theorem 4.2.4. Hence each nonzero nonunit is contained in at most finitely many idempotent maximal ideals. □

We say that R has *finite idempotent character* if each nonzero element is contained in at most finitely many idempotent maximal ideals and *finite unsteady character* if each nonzero element is contained in at most finitely many unsteady maximal ideals. By Corollary 4.2.9, a Prüfer domain that has weak factorization also has finite idempotent character. In the next theorem, we show that a Prüfer domain with weak factorization also has finite unsteady character. A consequence of this is that a Prüfer domain has weak factorization if and only if each unsteady maximal ideal has height one, each nonzero nonunit is contained in at most finitely many noninvertible maximal ideals and $(IR_M)^{-1} = I^{-1}R_M$ for each nonzero ideal I and each sharp maximal ideal M (see Theorem 4.2.12 below).

Recall from Sect. 2.5 that if $P \in \mathscr{S}$ is a prime ideal of a Prüfer domain R where $\mathscr{S}$ is a set of incomparable primes of R, then P is *relatively sharp* in $\mathscr{S}$ if it contains a finitely generated ideal that is contained in no other prime of the set $\mathscr{S}$ (or equivalently, R_P does not contain $\bigcap\{R_Q \mid Q \in \mathscr{S}\backslash\{P\}\}$). The set $\mathscr{S}$ is *relatively sharp* if each prime in $\mathscr{S}$ is relatively sharp in $\mathscr{S}$. We make use of these notions in the proof of our next theorem.

Theorem 4.2.10. *Let R be a Prüfer domain. If R has weak factorization, then R has finite unsteady character.*

Proof. By way of contradiction, assume there is an infinite set of unsteady maximal ideals $\mathscr{W} := \{M_\alpha \mid \alpha \in \mathscr{A}\}$ with a nonzero intersection $I := \bigcap_\alpha M_\alpha$. Each $M_\alpha \in \mathscr{W}$ has height one by Lemma 4.2.3(3). Also, by Corollary 4.2.9(1), no sharp prime contains I, so we may assume $\mathscr{W}$ is the complete set of minimal primes of I. Note that if $I^{-1} = R$, then we cannot have weak factorization, since in this case $I^v = R$, so it would be impossible to factor I as I^v times a finite product of maximal ideals. Thus we may further assume there is an element $t \in I^{-1}\backslash R$. Since R is Prüfer, the ideal $C := (R : (1,t))$ is an invertible ideal that contains I. Thus C is contained in no maximal ideal that does not contain I. Since I is a radical ideal and each of its minimal primes is maximal, C is a radical ideal as well with $\operatorname{Max}(R,C) \subseteq \operatorname{Max}(R,I) = \mathscr{W}$. It is easy to see that the set $\operatorname{Max}(R,C)$ must be infinite since no member of $\mathscr{W}$ is sharp and C is finitely generated (Theorem 2.3.11). Hence we may further assume that I is invertible.

If either $\mathscr{W}$ or an infinite subset of $\mathscr{W}$ is a relatively sharp set, then we have a contradiction by way of Corollary 4.2.9 and Theorem 2.5.9. Hence we may further assume no infinite subset of $\mathscr{W}$ is a relatively sharp set. Using this assumption we will arrive at a contradiction by constructing an infinite subset of $\mathscr{W}$ that is a relatively sharp set.

Let $M_\beta \in \mathscr{W}$ and let $q \in M_\beta\backslash I$. Then the ideal $E := qR+I$ is an invertible ideal that properly contains I. Moreover, for each $M_\alpha \in \mathscr{W}$, ER_{M_α} contains $IR_{M_\alpha} = M_\alpha R_{M_\alpha}$. Since E is invertible, $IEE^{-1} = I$. Thus the ideal $G := IE^{-1}$ is an invertible ideal of R that is contained in each maximal ideal of $\mathscr{W}$ that does not contain E and in no maximal ideal of $\mathscr{W}$ that contains E.

Suppose $M_1, M_2, \ldots, M_n \in \mathscr{W}$ are relatively sharp in $\mathscr{W}$, $n \geq 1$. Then for each i, there is a finitely generated ideal $J_i \subsetneq M_i$ such that no other member of $\mathscr{W}$ contains J_i. Moreover, we may assume each J_i contains I and $J_i + J_k = R$ for all $i \neq k$. Then the product $J := J_1 J_2 \cdots J_n$ contains I and is contained in each M_i but in no other member of $\mathscr{W}$. Since J is invertible, it follows that the ideal IJ^{-1} of R is contained in each maximal ideal of $\mathscr{W}$ except $M_1, M_2, \ldots, M_n$. Hence IJ^{-1} is the intersection of these ideals.

Since we have assumed at most finitely many members of $\mathscr{W}$ are relatively sharp in $\mathscr{W}$, we may further assume that no member of $\mathscr{W}$ is relatively sharp. Under this assumption, if B is a finitely generated ideal with $I \subsetneq B \subseteq M_\beta$ for some $M_\beta \in \mathscr{W}$, then B is contained in infinitely many members of $\mathscr{W}$ as is IB^{-1}, and no member of $\mathscr{W}$ contains both B and IB^{-1}.

With all of these assumptions, it is now relatively easy to construct a countably infinite subset of $\mathcal{W}$ that is relatively sharp and with this arrive at a contradiction.

We construct such a subset as follows.

Let $\mathcal{A}$ be a well-ordered index set for $\mathcal{W}$ and let α_1 be the smallest member of $\mathcal{A}$. Next set $M_1 := M_{\alpha_1}$, select an element $s_1 \in M_1 \backslash I$ and set $J_1 := I + s_1 R$. Then from the above, infinitely many members of $\mathcal{W}$ contain J_1 and infinitely many do not.

For M_2, let α_2 be the smallest $\alpha \in \mathcal{A}$ such that M_α does not contain J_1, then set $M_2 := M_{\alpha_2}$. Since infinitely many members of $\mathcal{W}$ do not contain J_1, there is an element $s_2 \in M_2 \backslash I J_1^{-1}$ such that s_1 and s_2 are comaximal but $s_1 s_2$ is not in I. Set $J_2 := I + s_2 R$. Then, clearly, $I \subsetneq J_1 \bigcap J_2 = J_1 J_2$. As above, infinitely many members of $\mathcal{W}$ do not contain $C_2 := J_1 J_2$.

Recursively, for $n \geq 3$, define ideals M_n, J_n and C_n as follows. Let α_n be the smallest $\alpha \in \mathcal{A}$ such that M_α does not contain C_{n-1}, then set $M_n := M_{\alpha_n}$. For J_n, there is an element $s_n \in M_n \backslash I C_{n-1}^{-1}$ such that s_n is comaximal with the ideal C_{n-1} but $s_n C_{n-1}$ is not contained in I. Set $J_n := I + s_n R$. Then $C_n := C_{n-1} J_n = C_{n-1} \bigcap J_n \supsetneq I$.

For $n \neq m$, the elements s_n and s_m are comaximal. Thus each M_m is relatively sharp in the set $\{M_n\}_{n=1}^{\infty}$, a contradiction to our assumption that no infinite subset of $\mathcal{W}$ is relatively sharp. Therefore it must be that no nonzero element is contained in infinitely many unsteady maximal ideals. □

The next result is a straightforward consequence of Corollary 4.2.9(3), Theorem 4.2.10, and Lemma 2.1.10.

Corollary 4.2.11. *Let R be a Prüfer domain. If R has weak factorization, then each nonzero nonunit is contained in at most finitely many noninvertible maximal ideals.*

Theorem 4.2.12. *The following statements are equivalent for a Prüfer domain R.*

(i) R has weak factorization.

(ii) (a) R is an aRTP-domain,
(b) each nonzero prime ideal is contained in a unique maximal ideal, and
(c) each nonzero nonunit is contained in at most finitely many noninvertible maximal ideals.

(iii) (a) Each steady maximal ideal is sharp,
(b) each nonzero nonmaximal prime is both sharp and contained in a unique maximal ideal, and
(c) each nonzero nonunit is contained in at most finitely many noninvertible maximal ideals.

(iv) (a) Each unsteady maximal ideal has height one,
(b) each nonzero ideal (or nonunit) is contained in at most finitely many noninvertible maximal ideals, and
(c) $(IR_M)^{-1} = I^{-1}R_M$ for each nonzero ideal I and each steady maximal ideal M.

(v) For each nonzero nondivisorial ideal I, there is a finite family of primes $\{P_1, P_2, \ldots, P_n\}$ such that $I = I^v P_1 P_2 \cdots P_n$.

(vi) *For each nonzero nondivisorial ideal I, there is a finite set of incomparable primes $\{Q_1, Q_2, \ldots, Q_m\}$ such that $I = I^v Q_1^{r_1} Q_2^{r_2} \cdots Q_m^{r_m}$ for some positive integers $r_1, r_2, \ldots, r_m$.*

Proof. For (i) implies (ii), simply apply Theorem 4.2.4 and Corollary 4.2.11. Also, the same argument used in the first part of the proof of Theorem 4.2.4 shows that (ii) and (iii) are equivalent.

To see that (iii) implies (iv), first apply Lemma 2.3.9 to see that each maximal ideal of height greater than one is sharp. Thus the only unsteady maximal ideals, if any, have height one. Also, by Theorem 2.5.4, $I^{-1}R_M = (IR_M)^{-1}$ for each nonzero ideal I and each steady maximal ideal M.

Next we show (iv) implies (i). Let I be a nonzero, nondivisorial ideal. By Theorem 2.5.4(2), if M is an invertible maximal ideal that contains I, then $IR_M = I^v R_M$. Thus there must be at least one noninvertible maximal ideal N that contains I and is such that $IR_N \neq I^v R_N$. Let $\{M_1, M_2, \ldots, M_n\}$ be the set of these maximal ideals. If M_i is idempotent, then $IR_{M_i} = I^v M_i R_{N_i}$ by Theorem 2.5.4(3). On the other hand, if M_i is locally principal (equivalently, not idempotent), then it has height one and there is a positive integer s_i such that $IR_{M_i} = I^v M_i^{s_i} R_{M_i}$ (see also the proof of Theorem 4.2.4((iii)⇒(i))). Checking locally shows that $I = I^v \prod_{i=1}^n M_i^{s_i}$ for some positive integers $s_1, s_2, \ldots, s_n$. Thus R has weak factorization.

Clearly, weak factorization implies the existence of the factorizations in (v) and (vi).

Since R is a Prüfer domain, if $Q \subsetneq P$ are distinct prime ideals, then $PQ = Q$. Hence in statement (v), if $P_i \subsetneq P_j$, then all occurrences of P_j can be removed from the factorization. It follows that (v) and (vi) are equivalent.

To complete the proof, we show that if each nonzero nondivisorial factors in the form given in (v), then R has weak factorization. For this, it suffices to show that R satisfies the criteria of Theorem 4.2.4(iii).

First, let P be a branched nonmaximal prime ideal. Then it is minimal over a finitely generated ideal I [34, Theorem 23.3(e)]. Let $J := IR_M \bigcap R$ where M is a maximal ideal that contains P. Then, clearly, $I \subseteq J \subseteq P$ and $\mathrm{Max}(R, J) = \mathrm{Max}(R, P)$. Note that $J \subsetneq P$; otherwise, $PR_M = JR_M = IR_M$ is a nonmaximal finitely generated prime in the valuation domain R_M, which is impossible. If P is not sharp, then $J \subsetneq P \subseteq P^v = J^v$ by Theorem 2.5.2(1). Now, consider a factorization $J = J^v P_1 P_2 \cdots P_n$ as in statement (v). In the valuation domain R_M, $JR_M = IR_M$ is principal, and therefore from this factorization of J, so are $J^v R_M$ and each $P_i R_M$. Hence either $P_i = M$ or $P_i R_M = R_M$. Since P is not maximal, it cannot contain any of the P_is. It follows that $P \supseteq J^v$ and hence $J^v = P$. Thus, as above, PR_M is a principal nonmaximal prime ideal of R_M, a contradiction. Therefore P is sharp.

If P is sharp but M is not the only maximal ideal that contains P, then J is not divisorial (Theorem 2.5.2(2)). Also, from the proof of Theorem 2.5.2(2), $J^v R_M = JR_{P'} = IR_{P'}$ where $P' \subsetneq M$ is the largest prime common to all maximal ideals that contain P. As above, factor J as $J = J^v P_1 P_2 \cdots P_n$.

We again have $J^v R_M$ and each $P_i R_M$ principal. However, since $IR_{P'} = J^v R_M$ is both a proper principal ideal of $R_{P'}$ and an ideal of R_M, this contradicts the fact that no proper principal ideal of a valuation domain can be an ideal in a proper overring. Thus having P sharp and in more than one maximal ideal is impossible. Therefore $J = J^v$, P is sharp and M is the only maximal ideal that contains P by Theorem 2.5.2(3).

Consequently, we have that each nonzero nonmaximal prime is sharp as is each maximal ideal of height greater than one (Lemma 2.3.9).

Thus according to Theorem 4.2.4, the only remaining case we need to consider (for sharpness) is that of a height one idempotent maximal ideal. Accordingly, let $M = M^2$ be a height one maximal ideal and let Q be a proper M-primary ideal. If Q^{-1} is a ring, then $Q^{-1} = M^{-1} = R$ by Lemma 2.3.15. Hence $Q^v = R$, and we have the (only possible) factorization $Q = Q^v M = M$, a contradiction. Thus Q^{-1} is not a ring, whence M is sharp by Theorem 2.3.17((i)⇔(iv)).

To complete the proof, we need only show that for each nonzero nondivisorial ideal B, the set of maximal ideals N for which $BR_N \neq B^v R_N$ is finite (criterion (iii)(c) of Theorem 4.2.4). Let B be a nonzero nondivisorial ideal. Then $B = B^v Q_1 Q_2 \cdots Q_m$ for some finite family of prime ideals $\{Q_1, Q_2, \ldots, Q_m\}$. Each Q_i is in a unique maximal ideal N_i and clearly $BR_N = B^v R_N$ for each maximal ideal N not in the family $\{N_1, N_2, \ldots, N_m\}$. Therefore R has weak factorization. □

We record the following simple consequence of Theorem 4.2.12 for ease of reference in Example 4.3.4 below.

Corollary 4.2.13. *Let R be a Prüfer domain with finite unsteady character. If each steady maximal ideal is invertible and each nonzero nonmaximal prime is sharp and contained in a unique maximal ideal, then R has weak factorization.*

One of the main concepts studied by Loper and Lucas in 2003 [58] is how far an almost Dedekind domain is from being Dedekind. For example, from what has already been observed in Sect. 3.2, an almost Dedekind domain R has sharp degree 2 if it is not Dedekind, but the intersection $R_2 := \bigcap R_M$ is Dedekind, where the intersection is taken over all maximal ideals M that are not invertible in R (in this case, they coincide with the dull maximal ideals of R, considered in Sect. 3.2). Note that R must have infinitely many invertible maximal ideals for this to happen. An example in [19] shows that an almost Dedekind domain with infinitely many noninvertible maximal ideals can have sharp degree 2 [19, Example 3.2]. Higher sharp degrees (including infinite ordinal degrees) can be defined recursively, as in Sect. 3.2. For example, an almost Dedekind domain R has sharp degree 3, if R_2 is not Dedekind and $R \subsetneq R_2 \subsetneq R_3$ with R_3 Dedekind where R_3 is the intersection of the localizations at the (nonempty set of) noninvertible maximal ideals of R_2. It turns out that an almost Dedekind domain that is not Dedekind has weak factorization if and only if it has sharp degree 2. More precisely, Theorem 4.1.5 essentially shows that an almost Dedekind domain with sharp degree 2 has weak factorization. In Proposition 4.2.14 below, we establish the converse, as well other ways to detect weak factorization in almost Dedekind domains.

Proposition 4.2.14. *Let R be an almost Dedekind domain that is not Dedekind. The following are equivalent.*

(i) R has weak factorization.
(ii) If $\{M_\alpha\}$ is a set of noninvertible maximal ideals such that $\bigcap_\alpha M_\alpha$ is nonzero, then the set $\{M_\alpha\}$ is finite.
(iii) Each noninvertible maximal ideal contains a finitely generated ideal that is contained in no other noninvertible maximal ideal.
(iv) If $\{M_\alpha\}$ is a nonempty set of noninvertible maximal ideals, then $\bigcap_\alpha R_{M_\alpha}$ is Dedekind.
(v) R has sharp degree 2.

Proof. Let $\{M_\gamma\}$ be the (complete) set of noninvertible maximal ideals of R. Then as observed above, R has sharp degree 2 if and only if $R_2 = \bigcap_\gamma R_{M_\gamma}$ is a Dedekind domain. Since an overring of a Dedekind domain is Dedekind, statements (iv) and (v) are equivalent. Moreover, if R_2 is Dedekind, then it is easy to see that each nonzero nonunit of R is contained in at most finitely many noninvertible maximal ideals. Thus (v) implies (i) by Theorem 4.2.12.

By Lemma 3.4.6, if (iii) holds, then each noninvertible maximal ideal becomes sharp in R_2 and from this and Theorem 2.2.1, we have that R has sharp degree 2 (so (iii) implies (v)).

Assume that R has weak factorization. Then by Theorem 4.2.12, each finitely generated nonzero ideal is contained in at most finitely many noninvertible maximal ideals. Thus an infinite intersection of noninvertible maximal ideals is zero, and we have that (i) implies (ii).

To see that (ii) implies (iii), let M be a noninvertible maximal ideal of R, and let a be a nonzero element of M. By (ii), a is contained in only finitely many noninvertible maximal ideals, say $M_1 = M, M_2, \ldots, M_n$. For each $1 < i \leq n$, pick an element $a_i \in M \setminus M_i$. Then the ideal $(a, a_2, \ldots, a_n)$ is contained in M and no other noninvertible maximal ideal of R. □

4.3 Overrings and Weak Factorization

Let R be a Prüfer domain with weak factorization. Also, let $\{M_\alpha\}$ be the set of unsteady maximal ideals, and assume that this set is nonempty. By Theorem 4.2.12, each M_α has height one, and each nonzero nonunit of R is contained in at most finitely many of the M_α. Let $T := \bigcap_\alpha R_{M_\alpha}$. By definition, sharp primes of R do not survive in T. Hence T is a one-dimensional Prüfer domain with finite character such that each localization is a rank one discrete valuation domain, that is, T is a Dedekind domain. Since an overring of a Dedekind domain is Dedekind, we have the following result.

Theorem 4.3.1. *Let R be a Prüfer domain. If R has weak factorization, then $\bigcap_\alpha R_{M_\alpha}$ is a Dedekind domain for each nonempty set of unsteady maximal ideals $\{M_\alpha\}$.*

Corollary 4.3.2. *Let R be a Prüfer domain with weak factorization. The following are equivalent for an overring T of R.*

(i) T has weak factorization.
(ii) T has finite idempotent character.
(iii) If $\{P_\alpha\}$ is a set of incomparable idempotent primes of R with a nonzero intersection, then at most finitely many P_α's extend to maximal ideals of T.

Proof. Let T be an overring of R, and let J be a nonzero ideal of T. Then $J = IT$ for some ideal I of R [34, Theorem 26.1(3)]. Also, if N is an idempotent maximal ideal of T, then $N = PT$ for some idempotent prime P of R. Thus statements (ii) and (iii) are equivalent.

Each unsteady maximal ideal of T is extended from an unsteady maximal ideal, and each sharp prime of R that survives in T is still sharp and contained in a unique maximal ideal of T. Note that an unsteady maximal ideal of R may extend to a steady maximal ideal of T, but in such a case the extension is an invertible height one maximal ideal (Lemma 4.2.3). Thus we have $J^{-1}T_N = (JT_N)^{-1}$ for each steady maximal ideal N of T (if N is extended from a steady maximal ideal M of R, then $I^{-1}R_M = (IR_M)^{-1}$ (Theorem 4.2.12), and so $J^{-1}T_N = (JT_N)^{-1}$; on the other hand, if N is extended from an unsteady maximal ideal M of R, then N is invertible in T with height one, in which case $J^{-1}T_N = (JT_N)^{-1}$ by Theorem 2.5.4). This shows that T satisfies condition (iv) of Theorem 4.2.12, so that T has weak factorization. Thus (ii) implies (i). Finally, (i) implies (ii) by Corollary 4.2.9(3). □

Corollary 4.3.3. *If R is Prüfer domain with weak factorization and no nonzero idempotent primes, then each overring has weak factorization.*

The next example shows that not all overrings of a Prüfer domain with weak factorization have weak factorization.

Example 4.3.4. Let $\{\mathrm{W}, \mathrm{X}, \mathrm{Y}_1, \mathrm{Y}_2, \dots, \mathrm{Z}_1, \mathrm{Z}_2, \dots\}$ be a countably infinite set of algebraically independent indeterminates over the field K and let $D := K[\mathrm{X}, \{\mathrm{Y}_n \mid n \geq 1\}, \{\mathrm{Z}_n^\alpha \mid n \geq 1, \alpha \in \mathbb{R}^+\}]$. For each $n \geq 1$, define a valuation v_n on the quotient field $F := K(\mathrm{X}, \{\mathrm{Y}_n \mid n \geq 1\}, \{\mathrm{Z}_n^\alpha \mid n \geq 1, \alpha \in \mathbb{R}^+\})$ of D with value group $\mathbb{R} \times \mathbb{Z}$ (lexicographically ordered) by first setting $\mathrm{v}_n(a) = (0,0)$ for all nonzero elements in K, $\mathrm{v}_n(\mathrm{X}) = (1,0)$, $\mathrm{v}_n(\mathrm{Y}_n) = (0,1)$, $\mathrm{v}_n(\mathrm{Z}_n^\alpha) = (\alpha, 0)$ and $\mathrm{v}_n(\mathrm{Y}_m) = \mathrm{v}_n(\mathrm{Z}_m^\alpha) = (0,0)$ for all $\alpha \in \mathbb{R}$ and $m \neq n$, and then extending v_n to F using "min." Let V_n be the corresponding valuation domain with quotient field F. By standard arguments, it can be shown that $\mathrm{Y}_n V_n$ is the maximal ideal of the two-dimensional valuation domain V_n. Let $R := \bigcap_n V_n(\mathrm{W})$, where $V_n(\mathrm{W})$ is the canonical (trivial) extension of V_n to the field of rational functions $F(\mathrm{W})$.

(1) R is a Bézout domain.
(2) For each n, the ideal $M_n := \mathrm{Y}_n R$ is an invertible height two maximal ideal.
(3) For each n, $P_n := \sqrt{\mathrm{Z}_n R}$ is a height one idempotent prime that is sharp and M_n is the only maximal ideal of R that contains P_n.
(4) Let J be the ideal generated by the set $\{\mathrm{X}/\mathrm{Z}_n \mid n \geq 1\}$. Then $M := \sqrt{J}$ is an unsteady height one maximal ideal.
(5) There are no other nonzero prime ideals in R.
(6) R has weak factorization. Moreover, each nonzero nondivisorial ideal I factors as $I^v M^k$ for some positive integer k.
(7) Let $\mathscr{S} := \{\mathrm{Y}_n^k \mid n \geq 1, k \geq 0\}$. The ring $R_{\mathscr{S}}$ is a one-dimensional Prüfer domain that does not have weak factorization.
(8) If $\mathscr{Q}$ is an infinite proper subset of $\{P_n\}_{n=1}^{\infty}$, then the ideal $H := \bigcap\{P_m \mid P_m \in \mathscr{Q}\}$ is a (nonzero) radical ideal such that $(H : H)$ does not have weak factorization.

Proof. For each n, let v_n^* denote the trivial extension of v_n to $F(\mathrm{W})$ (set the value of W to be 0, and extend to $F(\mathrm{W})$ using "min") [34, page 218]. The corresponding valuation domain is $V_n(\mathrm{W})$ [34, Propositions 18.7 and 33.1]. That R is a Bézout domain with $M_n = \mathrm{Y}_n R$ a maximal ideal, that $R_{M_n} = V_n(\mathrm{W})$ and that M_n has height two for each n follow from results on eab-operations and Kronecker function rings in [34, Chap. 32] (cf. also Halter–Koch [40, Theorem 2.2(2)]).

For each n, let $D_n := K[\mathrm{X}, \{\mathrm{Y}_m \mid 1 \leq m \leq n\}, \{\mathrm{Z}_m^{\alpha} \mid 1 \leq m \leq n, \alpha \in \mathbb{R}^+\}]$ and let $T_n := D_n[\mathrm{W}]$.

Let z be a nonzero element of the quotient field $F(\mathrm{W})$ of R. Then there is a pair of integers s and n such that z can be factored as a product $\mathrm{X}^s(g/f)$ where $g, f \in T_n \backslash \mathrm{X}T_n$. For each $k > n$, z is in $V_k(\mathrm{W})$ if and only if $s \geq 0$. Also, for $k > n$, z is a unit of $V_k(\mathrm{W})$ if and only if $s = 0$. Hence $z \in R$ implies $s \geq 0$, and $z \in \bigcap_n M_n$ implies $s > 0$.

By Theorem 2.5.10, $\bigcap_n M_n$ is the Jacobson radical of R and no maximal ideal other than one of the M_n's is sharp. Also, by Corollary 4.2.9, the only sharp primes that contain $\bigcap_n M_n$ are the M_n's. Since X is in each M_n, each maximal ideal contains X.

Let $t := a/b \in R\backslash M_n$ be a nonunit of R with $a, b \in D[\mathrm{W}]$ and let k be a positive integer greater than the largest power of Z_n that appears in a term of a. Then no "cancellation" can occur in the numerator of $t + \mathrm{Z}_n^k = (a + b\mathrm{Z}_n^k)/b$. In $V_n(\mathrm{W})$, $t + \mathrm{Z}_n$ is a unit since $t \notin M_n$. For $m \neq n$, $\mathrm{v}_m^*(a) \geq \mathrm{v}_m^*(b)$ and $\mathrm{v}_m^*(\mathrm{Z}_n^k) = \mathrm{v}_m(\mathrm{Z}_n^k) = (0, 0)$. From the definition of (the original) v_m, $\mathrm{v}_m^*(a + b\mathrm{Z}_n^k) = \mathrm{v}_m^*(b)$ and therefore $\mathrm{v}_m^*(t + \mathrm{Z}_n^k) = (0, 0)$. Hence $t + \mathrm{Z}_n^k$ is a unit of R.

Let N be a maximal ideal of R that does not contain some particular X/Z_n. Since $\mathrm{X} \in N$, then N must contain Z_n, and all positive powers of Z_n. From the argument in the previous paragraph, we must have $N = M_n$. This not only shows that M_n is the only maximal ideal that contains Z_n, it also shows that $P_n = \sqrt{\mathrm{Z}_n R}$ is a sharp prime and M_n is the only maximal ideal that contains P_n.

Let J be the ideal of R generated by the set $\{X/Z_n \mid n \geq 1\}$. Since each P_n contains the set $\{X/Z_m \mid m \neq n\}$, each element of J is in infinitely many P_n's. On the other hand, P_n does not contain X/Z_n. Thus no P_n contains J, nor does any M_n.

Let $E := K[W, \{Y_n \mid n \geq 1\}, \{Z_n^\alpha \mid n \geq 1, \alpha \in \mathbb{R}^+\}]$ and let $h \in E\backslash\{0\}$. Then there is an integer n such that $h \in E_n := K[W, \{Y_m \mid 1 \leq m \leq n\}, \{Z_m^\alpha \mid 1 \leq m \leq n, \alpha \in \mathbb{R}^+\}]$. Let $s_n := \sum_{m=1}^{n} X/Z_m$ and consider the element $h + s_n$. For $k > n$, h is a unit of $V_k(W)$ and s_n is a nonunit, and for $m \leq n$, $(0,0) = v_m(s_n) = v_m^*(s_n) = v_m^*(h + s_n)$. Thus $h + s_n$ is a unit of R. This implies that if M is a maximal ideal that contains J, then R_M contains L the quotient field of E. It follows that $R_M = L[X]_{(X)}$ is a discrete rank one valuation domain. We also have that M is the only prime that contains J. Hence M is height one, unsteady and the only other nonzero prime besides the M_n's and P_n's. Thus R has weak factorization by Corollary 4.2.13. By Theorem 2.5.4, if I is a nonzero nondivisorial ideal, then $IR_{M_n} = I^v R_{M_n}$ for each M_n and therefore $I = I^v M^k$ for some positive integer k.

Obviously, each M_n blows up in $R_{\mathscr{S}}$, but each P_n survives. Hence $R_{\mathscr{S}}$ is a one-dimensional Prüfer domain. As each element in M is in all but finitely many P_n's, the extension of M to $R_{\mathscr{S}}$ remains unsteady. Thus $R_{\mathscr{S}} = \bigcap_n R_{P_n}$. Let $B := \bigcap_n P_n R_{\mathscr{S}}$. Since X is in B, $P_n R_{\mathscr{S}}$ is minimal over B. As $P_n R_{P_n}$ is idempotent and $BR_{P_n} = P_n R_{P_n}$, $R_{P_n} \supseteq B(R_{\mathscr{S}} : B)R_{P_n} = P_n(R_{\mathscr{S}} : B)R_{P_n} = P_n R_{P_n}$. Hence $B(R_{\mathscr{S}} : B) = B$. Since $R_{\mathscr{S}} = \bigcap_n R_{P_n}$, $(R_{\mathscr{S}} : B) = R_{\mathscr{S}}$. Clearly, we cannot factor B as B^v $(= R_{\mathscr{S}})$ times a finite product of powers of maximal ideals. Hence $R_{\mathscr{S}}$ does not have weak factorization.

Finally, let $\mathscr{Q}$ be an infinite proper subset of $\{P_n\}_{n=1}^{\infty}$ and let $H := \bigcap\{P_m \mid P_m \in \mathscr{Q}\}$. As above $X \in H$. Thus by Corollary 4.2.9(2), $\mathscr{Q}$ is the complete set of sharp primes that are minimal over H. Also $H^{-1} = (H : H)$ as above since $HR_{P_m} = P_m R_{P_m}$ being idempotent implies $HH^{-1}R_{P_m} = HR_{P_m}$ for each $P_m \in \mathscr{Q}$. Hence each $P_m \in \mathscr{Q}$ extends to a (idempotent) maximal ideal of $(H : H)$ (Theorem 4.2.7). But this means that H is contained in infinitely many idempotent maximal ideals of $(H : H)$. Then Corollary 4.2.9(3) shows that $(H : H)$ does not have weak factorization. □

This example also shows that a ring of quotients of a Prüfer domain with weak factorization need not have weak factorization.

Theorem 4.3.5. *Let R be a Prüfer domain with weak factorization. If M is an unsteady maximal ideal, then there is a finitely generated ideal I and an infinite set of steady maximal ideals $\{M_\alpha\}$ each containing I, such that MT is the only unsteady maximal ideal of $T := \bigcap_\alpha R_{M_\alpha}$ and the only other maximal ideals of T are those of the form $M_\alpha T$.*

Proof. Let M be an unsteady maximal ideal and let J be a nonzero finitely generated ideal that is contained in M. Since R has finite unsteady character (Theorem 4.2.10), at most finitely many other unsteady maximal ideals contain J, say $M_1, M_2, \ldots, M_n$. For each i, there is an element $b_i \in M$ such that $b_i R + M_i = R$. Let $I := J + b_1 R + b_2 R + \cdots + b_n R$. Then M is the only unsteady maximal ideal that contains I.

Let $\{M_\alpha\}$ be the set of maximal ideals, other than M, that contain I. This set is infinite since M is not sharp. On the other hand, each M_α is steady and therefore sharp. By Lemma 2.5.1, the maximal ideals of $\Gamma(I)$ are the ideals of the form $M_\alpha\Gamma(I)$ and $M\Gamma(I)$. Since at most finitely many of the M_α are idempotent, $\Gamma(I)$ has weak factorization (Proposition 4.2.14). It also has infinitely many sharp maximal ideals and each maximal ideal contains the nonzero finitely generated ideal $I\Gamma(I)$. Hence $M\Gamma(I)$ must be the unique unsteady maximal ideal of $\Gamma(I)$. Moreover, $\Gamma(I) = \bigcap_\alpha R_{M_\alpha}$. □

Recall that for a nonzero ideal I of a domain R, $\mathrm{Min}(R, I)$ denotes the set of minimal primes of I (in R) and $\Phi(I) = \bigcap\{R_{P_\alpha} \mid P_\alpha \in \mathrm{Min}(R, I)\}$.

Lemma 4.3.6. *Let R be a Prüfer domain with weak factorization and let I be a nonzero ideal of R. Then $Max(\Phi(I)) = \{P\Phi(I) \mid P \in Min(R, I)\}$.*

Proof. Since R has weak factorization, the only nonzero primes that are not sharp are the unsteady maximal ideals, each of which has height one. If M is a maximal ideal that is minimal over I, then $M\Phi(I)$ is a maximal ideal of $\Phi(I)$. Let $P \subsetneq Q$ be primes with $P \in \mathrm{Min}(R, I)$. Since P is sharp, there is a finitely generated ideal $J \subseteq Q$ such that $P \subsetneq J \subseteq Q$ (Proposition 2.3.20). Obviously, no other minimal prime of I contains J. Hence R_Q does not contain $\Phi(I)$. It follows that $P\Phi(I)$ is a maximal ideal of $\Phi(I)$. □

Lemma 4.3.7. *Let R be a Prüfer domain with weak factorization and let I be a nonzero ideal that is not contained in the Jacobson radical of R. If N is a maximal ideal of $\Theta(I)$, then either $N \bigcap R$ is a maximal ideal of R that does not contain I or $N \bigcap R$ is an unsteady maximal ideal of R.*

Proof. Let P be a nonzero prime of R that is neither comaximal with I nor an unsteady maximal ideal of R. Since R has weak factorization, P is contained in a unique maximal ideal M and it is sharp (Theorem 4.2.4). Hence there is a (nonzero) finitely generated ideal $J \subseteq P$ such that M is the only maximal ideal that contains J. It follows that R_P does not contain $\Theta(I)$ $(= \bigcap\{R_Q \mid Q \in \mathrm{Max}(R)\backslash\mathrm{Max}(R, I)\})$. Since R is a Prüfer domain, $P\Theta(I) = \Theta(I)$. As each prime of $\Theta(I)$ is extended from a prime of R, if N is a maximal ideal of $\Theta(I)$, then either $N \bigcap R$ is a maximal ideal of R that does not contain I or it is an unsteady maximal ideal. □

Theorem 4.3.8. *Let R be a Prüfer domain with weak factorization, and let I be a nonzero ideal of R.*

(1) Both $\Gamma(I)$ and $\Theta(I)$ have weak factorization.
(2) $\Phi(I)$ has weak factorization if and only if at most finitely many minimal primes of I are idempotent.

Proof. By Lemma 2.5.1, $\mathrm{Max}(\Gamma(I)) = \{M\Gamma(I) \mid M \in \mathrm{Max}(R, I)\}$. As no nonzero ideal is contained in infinitely many idempotent maximal ideals (Corollary 4.2.9), the same occurs in $\Gamma(I)$. Hence $\Gamma(I)$ has finite idempotent character. Thus $\Gamma(I)$ has weak factorization by Corollary 4.3.2.

From Lemma 4.3.7, each maximal ideal of $\Theta(I)$ is extended from a maximal ideal of R. Hence $\Theta(I)$ also has finite idempotent character. Another application of Corollary 4.3.2 yields that $\Theta(I)$ has weak factorization.

For $\Phi(I)$, $\text{Max}(\Phi(I)) = \{P\Phi(I) \mid P \in \text{Min}(R, I)\}$ by Lemma 4.3.6. Thus $\Phi(I)$ has finite unsteady character if and only if at most finitely many minimal primes of I are idempotent. It follows that $\Phi(I)$ has weak factorization if and only if at most finitely many minimal primes of I are idempotent. □

Theorem 4.3.9. *If R is a Prüfer domain with weak factorization, then each overring has weak factorization if and only if there is no nonzero ideal with infinitely many idempotent minimal primes.*

Proof. From the previous theorem, if $I \neq (0)$ has infinitely many idempotent minimal primes, then $\Phi(I)$ does not have weak factorization. Conversely, suppose $T \supsetneq R$ is an overring that does not have weak factorization. Then by Corollary 4.3.2, there is a nonzero ideal B of T that is contained in infinitely many idempotent maximal ideals of T. It follows that the ring $\Gamma_T(B)$ has infinitely many idempotent maximal ideals, each of which contains B. Thus we may assume B is contained in each maximal ideal of T. Let M be a maximal ideal of T. Then $M = PR$ for some prime P of R. If M is idempotent, then P is a sharp prime of R. It follows that M is a sharp prime of T. By Corollary 2.5.7, M is minimal over the Jacobson radical of T. Hence the Jacobson radical has infinitely many idempotent minimal primes. □

4.4 Finite Divisorial Closure

If R is an h-local Prüfer domain and I is a nonzero nondivisorial ideal, then there is a finitely generated ideal $J \subseteq I^v$ such that $I + J = I^v$ [19, Proposition 2.10]. In the next lemma, we generalize this result by showing that a Prüfer domain with weak factorization has the same property.

Lemma 4.4.1. *Let R be a Prüfer domain with the weak factorization property. If I is a nonzero ideal of R, then there is a finitely generated ideal J such that $I + J = I^v$.*

Proof. If I is divisorial, there is nothing to prove. Hence we assume that I is not divisorial with factorization $I = I^v \prod_i N_i \prod_j M_j^{r_j}$. We may further assume that each N_i is steady with $I^v N_i \neq I^v$ and each M_j is unsteady with $r_j > 0$. At least one of the (finite) sets $\{N_i\}$ or $\{M_j\}$ is nonempty.

Since $M_j R_{M_j}$ is principal and has height one by Lemma 4.2.3, there is an element $a_j \in I^v$ such that $a_j R_{M_j} = I^v R_{M_j}$. Now, consider an N_i. Since it is steady, if it is locally principal, then it is invertible. But, in that case, $I^v N_i$ is a divisorial ideal that contains I and is properly contained in I^v, which is impossible. Thus it must be that N_i is idempotent with $I \subseteq I^v N_i \subsetneq I^v$. Hence $I^v R_{N_i}$ is principal by Lemma 2.5.3, and we may choose $b_i \in I^v$ such that $I^v R_{N_i} = b_i R_{N_i}$. Now, set $J := (a_1, \ldots, a_m, b_1, \ldots, b_n)$. To see that $I^v = I + J$ simply check locally.

By construction, we have $I + J \subseteq I^v$, $I^v R_{M_j} = a_j R_{M_j} \subseteq (I + J)R_{M_j}$ for each M_j, $I^v R_{N_i} = b_i R_{N_i} \subseteq (I + J)R_{N_i}$ for each N_i, and $I^v R_N = IR_N \subseteq (I + J)R_N$ for all other maximal ideals N (if any). Hence $I^v = I + J$. □

We say that R has the *finite divisorial closure property* if, for each nondivisorial ideal $I \neq (0)$, there is a finitely generated ideal J such that $I^v = I + J$. A Prüfer domain with the finite divisorial closure property need not have weak factorization. For example, in the ring of entire functions the only divisorial ideals are the principal ones, so that this Prüfer domain has the finite divisorial closure property trivially. However, the primes reverse roles from what occurs with weak factorization—the only sharp primes are the height one invertible maximal ideals, all other primes have infinite height, and none of these is sharp, but each is contained in a unique maximal ideal. (For the properties of the ring of entire functions mentioned above see, for example, [47], [34, Pages 146–148 and Exercise 19, page 256] and [24, Sect. 8.1].)

In Theorem 4.4.7, we combine the finite divisorial closure property with another to obtain yet another characterization for Prüfer domains with weak factorization. In Theorem 4.4.8, we do the same for h-local Prüfer domains.

Lemma 4.4.2. *Let R be a Prüfer domain and let I be a nonzero nondivisorial ideal. If there is a finitely generated ideal J such that $I + J = I^v$, then for each maximal ideal M containing I with $IR_M \neq I^v R_M$, $I^v R_M$ is principal.*

Proof. Assume that $I^v = I + J$ for some finitely generated ideal J, and let M be a maximal ideal such that $IR_M \neq I^v R_M$. Since R_M is a valuation domain, $I^v R_M = JR_M$ is principal. □

Theorem 4.4.3. *Let R be a Prüfer domain with the finite divisorial closure property.*

(1) If P is a nonzero nondivisorial prime, then $P^v = R$.
(2) If P is a nonzero divisorial prime, then P is sharp and contained in a unique maximal ideal.

Proof. Let P be a nonzero prime ideal of R. We may assume that P is not maximal since both parts of the theorem hold trivially if P is maximal (Lemma 2.1.1(1) and Remark 2.1.2(1)).

For (1), we assume that P is not divisorial. In this case, there is a finitely generated ideal A such that $P^v = P + A \supsetneq P$. Then $P^{-1} = (P : P)$ (Theorem 2.3.2(2)), and, since R is integrally closed, $P^{-1} = (P^v)^{-1} = (\sqrt{P^v})^{-1}$ by Remark 2.3.3. Thus $(\sqrt{P^v})P^{-1} = (\sqrt{P^v})(\sqrt{P^v})^{-1} \subseteq R$, whence $(\sqrt{P^v}) \subseteq P^v$, and we have that P^v is a radical ideal. Now, by way of contradiction, suppose that $P^v \neq R$. If M is a maximal ideal that contains P^v, then $P^v R_M$ is a prime ideal of the valuation domain R_M that properly contains PR_M. Hence $P^v R_M = (P + A)R_M = AR_M$ is a principal prime ideal, and we must have $P^v R_M = MR_M$. It follows that each minimal prime of P^v is a maximal ideal of R and therefore $\Gamma(P^v) = \Phi(P^v)$. But in this case, $P^{-1} = (P^v)^{-1} = \Gamma(P^v) \bigcap \Theta(P^v) = R$ (Theorem 2.3.2(1, b)), whence $P^v = R$, the desired contradiction.

For (2), first assume P is both divisorial and branched. Then there is a proper P-primary ideal Q. If P is not sharp, then $\Theta(P) = (P : P) \subseteq P^{-1} \subseteq Q^{-1} = \Theta(P)$ (Corollary 2.3.18). Hence $Q^v = P^v = P$. Thus we have a finitely generated ideal B such that $P = Q^v = Q + B$. As $Q \neq Q^v$, there is a maximal ideal M that (properly) contains P with $QR_M \subsetneq PR_M = Q^v R_M = BR_M$, which is impossible since P is not maximal. Thus it must be that P is sharp.

Continuing with the assumption that P is both divisorial and branched, and now sharp as well, let A be a finitely generated ideal with radical P (Proposition 2.3.10). Also, let M be a maximal ideal that contains P and let $C := AR_M \bigcap R$. Then by Theorem 2.5.2, $C^v = A(P' : P')$, where P' is the largest prime common to all maximal ideals that contain P. If M is not the only maximal ideal that contains P, then C is not divisorial (Theorem 2.5.2 again), and so there is a finitely generated ideal $J \not\subseteq C$ such that $C^v = C + J$. We may assume that $A \subseteq J$, which implies that $CR_M = AR_M \subseteq JR_M$. Hence $JR_M = C^v R_M = A(P' : P')R_M = AR_{P'}$, the latter equality following from the fact that P' is a maximal ideal of $(P' : P') = \Theta(P) \bigcap R_{P'}$ (Lemma 2.3.9 and Theorem 2.3.2(2)(b)). But, since P' is properly contained in M, $AR_{P'}$ cannot be an invertible ideal of R_M. Thus M must be the only maximal ideal that contains P.

The only case left is when P is a (nonmaximal) prime ideal that is both divisorial and unbranched. In this case, P contains a nonzero branched prime P_0 which, from (1), cannot be nondivisorial. Thus P_0 is divisorial and branched, and therefore by the above, it is sharp and contained in a unique maximal ideal. It follows that this same maximal ideal is the only one that contains P. Hence P is sharp by Lemma 2.3.9. □

Corollary 4.4.4. *Let R be a Prüfer domain. If R has the finite divisorial closure property, then the following statements are equivalent.*

(i) Each nonzero nonmaximal prime is sharp and contained in a unique maximal ideal, and each maximal ideal of height greater than one is sharp.
(ii) Each nonzero nonmaximal branched prime is sharp (i.e., R is a wTPP-domain by Theorem 2.4.17).
(iii) Each nonzero nonmaximal branched prime is divisorial.

Proof. Obviously, (i) implies (ii). Also, (ii) implies (iii) since a nonmaximal sharp prime in a Prüfer domain must be divisorial (Corollary 2.3.21).

Assume that R has the finite divisorial closure property. If each nonzero nonmaximal branched prime is divisorial, then each is sharp and contained in a unique maximal ideal by Theorem 4.4.3(2). A maximal ideal of height greater than one contains a nonzero nonmaximal branched prime as does an unbranched (nonzero) prime. But such a branched prime is contained in a unique maximal ideal. Thus each maximal ideal of height greater than one is sharp as is each unbranched prime (Lemma 2.3.9). □

From Theorem 2.4.18 and the previous corollary, we immediately deduce the following.

Corollary 4.4.5. *Let R be a Prüfer domain. If R is an aRTP-domain and has the finite divisorial closure property, then each nonzero prime is contained in a unique maximal ideal.*

We do not know whether the conclusion of Corollary 4.4.5 can be strengthened to "R has the weak factorization property" or not.

Lemma 4.4.6. *Let I be a nonzero nondivisorial ideal in a Prüfer domain R. If there is a finitely generated ideal $J \subseteq I^v$ and a finite set of maximal ideals $\{M_1, M_2, \ldots, M_n\}$ such that $I^v = I + J$ and $J \prod M_i^{r_i} \subseteq I$ for some positive integers $r_1, r_2, \ldots, r_n$, then $I = I^v \prod M_i^{s_i}$ for some nonnegative integers $s_1, s_2, \ldots, s_n$ with $s_i \leq r_i$ for each i.*

Proof. Assume that there is a finitely generated ideal $J \subseteq I^v$ and a finite set of maximal ideals $\{M_1, M_2, \ldots, M_n\}$ such that $I^v = I + J$ and $J \prod M_i^{r_i} \subseteq I$ for some positive integers $r_1, r_2, \ldots, r_n$. For a maximal ideal M outside the set $\{M_1, M_2, \ldots, M_n\}$, we have $JR_M \subseteq IR_M \subseteq I^v R_M = (I + J)R_M$. Thus $IR_M = I^v R_M$.

We may divide the set $\{M_1, M_2, \ldots, M_n\}$ into three disjoint subsets: $\mathscr{A}_1(I) := \{M_i \mid IR_{M_i} = I^v R_{M_i}\}$, $\mathscr{A}_2(I) := \{M_i \mid IR_{M_i} \neq I^v R_{M_i}$ and M_i is idempotent$\}$ and $\mathscr{A}_3(I) := \{M_i \mid IR_{M_i} \neq I^v R_{M_i}$ and M_i is locally principal$\}$. While $\mathscr{A}_1(I)$ may be empty, at least one of $\mathscr{A}_2(I)$ and $\mathscr{A}_3(I)$ must be nonempty since I is not divisorial. Note that $\prod M_i^{r_i} R_{M_j} = M_j^{r_j} R_{M_j}$ for each j. Hence we have $JM_j^{r_j} R_{M_j} \subseteq IR_{M_j}$. Also, since each R_{M_j} is a valuation domain, $IR_{M_j} \subsetneq I^v R_{M_j} = JR_{M_j}$ for each $M_j \in \mathscr{A}_2(I) \bigcup \mathscr{A}_3(I)$.

The maximal ideals in $\mathscr{A}_1(I)$ and $\mathscr{A}_2(I)$ are quite easy to deal with. For $M_i \in \mathscr{A}_1(I)$, we set $s_i = 0$. For $M_j \in \mathscr{A}_2(I)$, $M_j^r = M_j$ for each positive integer r. Also, there can be no ideals properly between $JM_j R_{M_j} = I^v M_j R_{M_j}$ and $JR_{M_j} = I^v R_{M_j}$ for $M_j \in \mathscr{A}_2(I)$. Since $I^v M_j R_{M_j} = JM_j R_{M_j} \subseteq IR_{M_j} \subsetneq I^v R_{M_j}$, we have $IR_{M_j} = I^v M_j R_{M_j}$ and we may set $s_j = 1$.

Finally, for $M_k \in \mathscr{A}_3(I)$, the only ideals between $JM_k^{r_k} R_{M_k} = I^v M_k^{r_k} R_{M_k}$ and $JR_{M_k} = I^v R_{M_k}$ are the ideals of the form $JM_k^s R_{M_k} = I^v M_k^s R_{M_k}$ for each nonnegative integer $s \leq r_k$. As observed above, one such ideal is IR_{M_k}. Thus $IR_{M_k} = I^v M_k^{s_k} R_{M_k}$ for some positive integer $s_k \leq r_k$. Checking locally at each maximal ideal now yields $I = I^v \prod M_i^{s_i}$ for the nonnegative integers $s_1, s_2, \ldots, s_n$ (in each case with $s_i \leq r_i$). □

Theorem 4.4.7. *Let R be a Prüfer domain. Then R has weak factorization if and only if for each nonzero nondivisorial ideal I, there is an invertible ideal $J \subseteq I^v$, a finite set of maximal ideals $\{M_1, M_2, \ldots, M_n\}$ and positive integers $r_1, r_2, \ldots, r_n$ such that $I^v = I + J$ and $J \prod M_i^{r_i} \subseteq I$.*

Proof. If I is a nondivisorial ideal and R has weak factorization, then there is a finite set of maximal ideals $\{M_1, M_2, \ldots, M_n\}$ and positive integers $r_1, r_2, \ldots, r_n$ such that $I = I^v \prod M_i^{r_i}$. By Lemma 4.4.1, there is an invertible ideal $J \subseteq I^v$ such that $I^v = I + J$. Obviously, we also have $J \prod M_i^{r_i} \subseteq I$.

For the converse, simply apply Lemma 4.4.6. □

As we recalled in Theorem 4.2.1, the equivalence of (1) and (6) in the following theorem originally appeared in [19] as part of Theorem 1.13. Using Theorem 4.2.4, we will give an alternate proof that (6) implies ((5) implies) (1).

Theorem 4.4.8. *The following statements are equivalent for a Prüfer domain R.*

(i) R is h-local.
(ii) For each nonzero nondivisorial ideal I, there is a finite set of distinct maximal ideals $\{M_1, M_2, \dots, M_k\}$ such that $I = I^v M_1 M_2 \cdots M_k$.
(iii) For each nonzero nondivisorial ideal I, there is a finite set of incomparable primes $\{Q_1, Q_2, \dots, Q_m\}$ such that $I = I^v Q_1 Q_2 \cdots Q_m$.
(iv) For each nonzero nondivisorial ideal I, there is a finite set of distinct prime ideals $\{P_1, P_2, \dots, P_n\}$ such that $I = I^v P_1 P_2 \cdots P_n$.
(v) R has the weak factorization property and no unsteady maximal ideals.
(vi) R has both the weak factorization property and finite character.
(vii) R has the weak factorization property, and each maximal ideal of R is sharp.
(viii) R has the weak factorization property, and each nonzero nonunit is contained in only finitely many invertible maximal ideals.
(ix) R has both finite character and the finite divisorial closure property.
(x) R has both the radical trace property and the finite divisorial closure property.
(xi) For each nonzero nondivisorial ideal I of R, there is a finite nonempty set of maximal ideals $\{M_1, M_2, \dots, M_n\}$ and a finitely generated ideal $J \subseteq I^v$ such that $I^v = I + J$ and $J \prod M_i \subseteq I$.

Proof. We establish the following sets of implications: (vi)⇔ (viii); (iii)⇔ (iv); (i)⇒ (xi)⇒ (ii)⇒ (iii)⇒ (v)⇒ (i); (i)⇒ (ix)⇒ (x)⇒ (i); and (i)⇒ (vi)⇒ (vii)⇒ (v). We note that (i) ⇒ (ii) is clear from Theorem 4.1.2.

It is clear that (vi) implies (viii). Also, we have that (viii) implies (vi) since weak factorization implies that each nonzero nonunit of R is contained in only finitely many noninvertible maximal ideals (Theorem 4.2.12). Several of the other implications are also easy to deal with. The equivalence of (iii) and (iv) follows from the fact that, checking locally, $QP = Q$ if $Q \subsetneq P$ are primes (in a Prüfer domain). Thus a factorization as in (iv) can simply be reduced to one with incomparable primes.

Next, we establish the series of implications (i)⇒ (xi)⇒ (ii)⇒ (iii)⇒ (v)⇒ (i). It is clear that (ii) implies (iii). For (xi) implies (ii), just apply Lemma 4.4.6. To see that (iii) implies (v), assume that (iii) holds. Then R has weak factorization by Theorem 4.2.12 ((vi)⇒ (i)). In order to prove (v), it remains to show that R has no unsteady maximal ideals. For this, suppose that M is a maximal ideal that is locally principal. Then $M^2 \neq M$. If M is not invertible, then $(R : M^2) = ((R : M) : M) = (R : M) = R$ (Remark 2.1.2(1)), in which case by (iii) we must have $M^2 = (M^2)^v M = R \cdot M = M$, a contradiction. Thus we must have M invertible. Hence (iii) implies (v).

Next, we show that (v) implies (i). Hence we assume that R has weak factorization and no unsteady maximal ideals. By Theorem 4.2.12, each nonzero prime is contained in a unique maximal ideal, and, since there are no unsteady maximal ideals, each nonzero (branched) prime ideal is sharp. Thus R is an RTP-domain by Theorem 2.4.10. Therefore R is h-local by Theorem 2.4.12.

To complete the sequence of implications (i)⇒(xi)⇒ (ii)⇒ (iii)⇒ (v)⇒ (i), we show (i)⇒ (xi). Assume that R is h-local, and let I be a nondivisorial ideal of R. By Lemma 4.4.1 and the implications (i)⇒ (ii)⇒ (iii)⇒ (v) (or, directly, by [19, Proposition 2.10]), there is a finitely generated ideal J such that $I^v = I + J$. Since R also has very strong factorization (Theorem 4.1.2), there is a finite set of maximal ideals $\{M_1, M_2, \ldots, M_n\}$ with $I = I^v \prod M_i = (I + J) \prod M_i$. Of course, this yields $J \prod M_i \subseteq I$, and we have that (i) implies (xi), completing the set. This also gives (i) implies (ix) since an h-local domain has finite character by definition.

To complete the set (i)⇒ (ix)⇒ (x)⇒ (i), we need (ix)⇒ (x) and (x)⇒ (i). If R has finite character, then it has the radical trace property by Theorem 2.4.11(2) and Lemma 2.3.4(2). Hence (ix) implies (x). Since a RTP-domain is clearly an aRTP-domain, (x) implies (i) by Corollary 4.4.5 and Theorem 2.4.12.

Finally, we consider the set (i)⇒ (vi)⇒ (vii)⇒ (v). By definition, if R is h-local, then it has finite character. Also it has very strong, hence weak, factorization by Theorem 4.1.2. Thus we have (i) implies (vi). An unsteady maximal ideal is not sharp. Thus (vii) implies (v).

Assume (vi). If M is a maximal ideal of R, then using finite character (and prime avoidance), it is easy to produce a finitely generated ideal I such that $\sqrt{I} = M$. Hence M is sharp (Proposition 2.3.10). Also, recall that a sharp maximal ideal must be steady by Lemma 2.4.19. Therefore (vi) implies (vii). □

By Theorem 4.1.2((i)⇔(ii)), the equivalence of (i) and (ii) in Theorem 4.4.8 means that, *for Prüfer domains, very strong factorization and strong factorization are equivalent.* Without the Prüfer assumption, we still have the following general result.

Theorem 4.4.9. *Let R be an integral domain. If R has strong factorization, then each nonzero finitely generated ideal of R is divisorial (equivalently, each nonzero ideal is a t-ideal).*

Proof. Suppose that R has strong factorization. We first observe, by checking locally, that if A is a nonzero ideal of R with factorization $A = A^v N_1 N_2 \cdots N_k$ with, say, $AR_{N_1} = A^v R_{N_1}$, then N_1 can be omitted from the factorization. Now, by way of contradiction, assume that I is a nonzero finitely generated ideal of R which is not divisorial and write $I = I^v \prod M_i$ for distinct maximal ideals $M_1, M_2, \ldots, M_n$ (and $n \geq 1$). We may assume no M_i can be omitted. Let $Q := \prod M_i$ and consider the ideal IQ. Calculating the divisorial closure, we have $(IQ)^v = (I^v Q)^v = I^v$. If IQ is divisorial, then $I^v = IQ \subseteq I$ and so I is also divisorial, a contradiction. Thus IQ is not divisorial and must therefore have a factorization as I^v times a finite product of distinct maximal ideals. Clearly, $IQR_M = IR_M = I^v R_M$ for each maximal ideal M outside the set $\{M_1, M_2, \ldots, M_n\}$. Hence the only possible way to factor IQ is as $IQ = I^v Q_1$, where Q_1 is a product of a subset of the M_i. However, this yields $I \subseteq IQ$, which contradicts Nakayama's Lemma. Hence each nonzero finitely generated ideal of R is divisorial. The parenthetical statement follows easily from the definition of the t-operation. □

One of many characterizations of Prüfer domains is that an integrally closed domain is Prüfer if and only if each nonzero finitely generated ideal is divisorial (see, for example, [34, Proposition 34.12]). Also, a celebrated result of Heinzer states that in an integrally closed domain R, each nonzero ideal is divisorial if and only if R is Prüfer h-local with all maximal ideals invertible (Theorem 2.1.6). The next result shows that, in the class of integrally closed integral domains, Prüfer domains with strong factorization lie in between the domains in which all nonzero ideals are divisorial and those in which all nonzero finitely generated ideals are divisorial.

Corollary 4.4.10. *Let R be an integrally closed domain. Then R has strong factorization if and only if R is an h-local Prüfer domain.*

Proof. Simply apply [34, Proposition 34.12] and Theorems 4.4.9 and 4.4.8. □

We close this section by showing that, in the integrally closed case, a domain with weak factorization must also be Prüfer. We need a preliminary result.

Lemma 4.4.11. *Let R be a domain with weak factorization. Then R_M is a valuation domain for each nondivisorial maximal ideal M of R.*

For a nonzero fractional ideal J of R, there is a nonzero element $r \in R$ such that rJ is an (integral) ideal of R. Moreover, $(rJ)^v = rJ^v$ so that J is divisorial if and only if rJ is divisorial. Hence we may easily extend weak (and strong) factorization to fractional ideals.

Proof. Let M be a nondivisorial maximal ideal, and assume $x \in K \setminus R_M$ where K is the quotient field of R. We shall show that $x^{-1} \in R_M$. Consider the fractional ideal $M + Rx$. Since M is nondivisorial, we have $(M + Rx)^v = (M^v + Rx)^v = (R + Rx)^v$. If $M + Rx$ is divisorial, this yields $1 \in (R + Rx)^v = M + Rx$, and we can write $1 = m + rx$ with $m \in M, r \in R$. In this case, we have $x^{-1} = r(1-m)^{-1} \in R_M$, as desired. Hence we assume that $M + Rx$ is not divisorial, in which case we have a factorization $M+Rx = (M+Rx)^v Q = (R+Rx)^v Q$, where Q is a product of (not necessarily distinct) maximal ideals. Suppose that $R + Rx$ is divisorial. Then $M + Rx = (R + Rx)Q$. If $Q \subseteq M$, then $M + Rx \subseteq M + Mx$, and we can write $x = a + bx$ with $a, b \in M$, whence $x = a(1-b)^{-1} \in R_M$, a contradiction. On the other hand, if $Q \nsubseteq M$, we have $MR_M + R_M x = R_M + R_M x$, which yields $1 \in MR_M + R_M x$, and we obtain $x^{-1} \in R_M$, as above.

It remains to consider the case where $R+Rx$ is not divisorial (and $M+Rx$ is also not divisorial). Recall that we have $M+Rx = (R+Rx)^v Q$. Since (we are assuming that) $R + Rx$ is not divisorial, we have a factorization $R + Rx = (R + Rx)^v Q'$, where Q' is a product of maximal ideals. Let $I := (R + Rx)^v$. Locally, we have $MR_M + R_M x = IR_M(MR_M)^i$ and $R_M + R_M x = IR_M(MR_M)^j$, for some nonnegative integers i, j. If $i \leq j$, then $MR_M + R_M x \supseteq R_M + R_M x$, and we obtain

$x^{-1} \in R_M$ as above. If $i > j$, then $MR_M + R_M x = IR_M(MR_M)^j(MR_M)^{i-j} = (R_M + MR_M x)(MR_M)^{i-j} \subseteq MR_M + MR_M x$, and, as before, we obtain the contradiction $x \in R_M$. This completes the proof. □

Theorem 4.4.12. *If R is an integrally closed domain with weak factorization, then R is a Prüfer domain.*

Proof. As in Theorem 4.4.9, we show that each nonzero finitely generated ideal is divisorial. Suppose, on the contrary, that I is a nonzero finitely generated ideal which is not divisorial. Then we may write $I = I^v Q$ with Q a product of (not necessarily distinct) maximal ideals. We still have $(IQ)^v = (I^v Q)^v = I^v$ as in the proof of Theorem 4.4.9. This yields $Q^{-1}I^v = Q^{-1}(QI)^v \subseteq (Q^{-1}QI)^v \subseteq I^v$. Hence $Q^{-1}I = Q^{-1}I^v Q \subseteq I^v Q = I$. Since I is finitely generated and R is integrally closed, then $R \subseteq Q^{-1} \subseteq (I : I) = R$ [34, Proposition 34.7], thus we must have $Q^{-1} = R$. It follows that each factor M of Q must be a nondivisorial maximal ideal. By Lemma 4.4.11, R_M is a valuation domain for each such M. Hence IR_M is a principal ideal of R_M. Thus $I^v R_M = (I^{-1})^{-1}R_M \subseteq (I^{-1}R_M)^{-1} = (IR_M)^v = IR_M$, and we have $IR_M = I^v R_M$. Since this equality obviously holds for any maximal ideal which is not a factor of Q, we obtain the contradiction that $I = I^v$. □

Chapter 5
Pseudo-Dedekind and Strong Pseudo-Dedekind Factorization

Abstract The present chapter is devoted to the study of integral domains having two other kinds of ideal factorization. An integral domain is said to have strong pseudo-Dedekind factorization if each proper ideal can be factored as the product of an invertible ideal (possibly equal to the ring) and a finite product of pairwise comaximal prime ideals with at least one prime in the product. On the other hand, an integral domain is said to have pseudo-Dedekind factorization if each nonzero noninvertible ideal can be factored as the product of an invertible ideal (which might be equal to the ring) and finitely many pairwise comaximal primes. We observe that an integral domain with pseudo-Dedekind factorization has strong factorization (Sect. 4.1) and an integrally closed domain with pseudo-Dedekind factorization is an h-local Prüfer domain. Nonintegrally closed local domains with pseudo-Dedekind factorization are fully described in terms of pullbacks of valuation domains. Several characterizations of integral domains with strong pseudo-Dedekind factorization are also given. In particular, we show that an integral domain has strong pseudo-Dedekind factorization if and only if it is an h-local generalized Dedekind domain. Finally, we investigate the ascent and descent of several types of ideal factorizations from an integral domain R to the Nagata ring $R(X)$ and vice versa.

5.1 Pseudo-Dedekind Factorization

In 2000, Olberding introduced the notion of a *ZPUI-ring* (or a *Zerlegung Prim- und Umkehrbaridealen ring*) as a ring such that each proper ideal can be factored as the product of an invertible ideal (possibly equal to the ring) and a finite product of prime ideals with at least one prime in the product [68]. We introduce two variations on this concept. We say that R has *pseudo-Dedekind factorization* if for each nonzero noninvertible ideal I, there is an invertible ideal B (which might be R) and finitely many pairwise comaximal primes $P_1, P_2, \ldots, P_n$ such that $I = BP_1P_2 \cdots P_n$ (the requirement that $n > 0$ "comes for free"). If, in addition, each invertible ideal J has a factorization of the form $J = CQ_1Q_2 \cdots Q_m$

M. Fontana et al., *Factoring Ideals in Integral Domains*, Lecture Notes of the Unione Matematica Italiana 14, DOI 10.1007/978-3-642-31712-5_5,

with C invertible (possibly with $C = R$) and pairwise comaximal prime ideals $Q_1, Q_2, \ldots, Q_m$ with $m \geq 1$, then R has *strong pseudo-Dedekind factorization*. Thus a domain that has strong pseudo-Dedekind factorization is a ZPUI-domain. In 1998, Olberding [67, Theorem 5.2] proved that a Prüfer domain R has (in our terminology) strong pseudo-Dedekind factorization if and only if R is an h-local generalized Dedekind domain. In subsequent work [68] and [70], he has been able to eliminate the assumption that R is a Prüfer domain and prove further that one only needs a factorization in terms of a finitely generated ideal and a nonempty finite product of not necessarily comaximal primes. Hence a ZPUI-domain is the same as a domain with strong pseudo-Dedekind factorization. Note that [70] corrects some gaps/errors that appear in [68]. We provide a simple example of a domain with pseudo-Dedekind factorization that is not integrally closed (and hence does not have strong pseudo-Dedekind factorization) after the proof of Corollary 5.2.5.

Recall that a valuation domain V is strongly discrete if no nonzero prime ideal is idempotent (equivalently if PV_P is principal for each nonzero prime ideal P). In 1987, Anderson [1] proved a result that now can be stated as follows: *A strongly discrete valuation domain is a ZPUI-domain*. More precisely,

Theorem 5.1.1. (Anderson [1, Theorem 2]). *Let V be a valuation domain.*

(1) If I is a nonzero ideal of V such that $I(V : I) = P$ is a branched prime of V, then $I = xQ$ for some $x \in V$ and some P-primary ideal Q.
(2) If each nonzero prime ideal of V is branched, then each nonzero noninvertible ideal is the product of a principal ideal and a primary ideal.
(3) If V is a strongly discrete valuation domain, then each nonzero noninvertible ideal I is the product of a principal ideal and a power of a prime ideal (more precisely, $I = xP^n$, for some $n \geq 1$ and some $x \in V$, where $P := I(V : I)$).

Both (1) and (2) are directly from [1, Theorem 2]. For (3), first recall that if I is a nonzero noninvertible ideal of a valuation domain V, then II^{-1} is a prime ideal of V (Proposition 2.4.1). If, in addition, $P := II^{-1}$ is such that PV_P is principal, then not only is P branched, but $\{P^n\}$ is the complete set of P-primary ideals [34, Theorem 17.3 (b)]. Statement (3) now follows easily from (1).

Remark 5.1.2. (1) We note the following with respect to statements (1) and (2) of Theorem 5.1.1.

(a) There exists a valuation domain with an unbranched maximal ideal and a nonzero noninvertible ideal I such that I is not the product of a principal ideal and a primary ideal [1, Example].
(b) There exists a valuation domain with an unbranched maximal ideal such that each noninvertible ideal is the product of a principal ideal and a prime ideal [15, Theorem 7].

(2) If I is a nonzero noninvertible nondivisorial ideal of a valuation domain V with maximal ideal M, then it is known that M is not finitely generated (otherwise, every nonzero ideal of V would be divisorial) and $I = xM$ for some $x \in V$ by Lemma 2.1.1(5) (or, by [44, Lemma 5.2]

and [6, Lemma 4.2]; see also [34, Exercise 12, page 431]). In this situation, $I(V:I) = xM(V:xM) = M(V:M) = M$, where the last equality holds because the maximal ideal M is not invertible (Lemma 2.1.1(1)). Therefore no matter whether M is branched or unbranched, we have a factorization, as in Theorem 5.1.1(1), for the nonzero noninvertible nondivisorial ideals of a valuation domain.

Lemma 5.1.3. *Let R be an integral domain with pseudo-Dedekind factorization.*

(1) R_S has pseudo-Dedekind factorization for each multiplicative set S of R.
(2) For each prime P of R, R/P has pseudo-Dedekind factorization.

Proof. If J is a noninvertible ideal of R_S, then $J = IR_S$ for some noninvertible ideal I of R. Factoring $I = BP_1P_2\cdots P_n$ with B invertible and the P_i pairwise comaximal primes (each with empty intersection with S), $J = BP_1P_2\cdots P_nR_S$ with BR_S invertible and $P_1R_S, P_2R_S, \ldots, P_nR_S$ pairwise comaximal primes of R_S. It follows that R_P has pseudo-Dedekind factorization for each prime P.

Let A be a nonzero noninvertible ideal of the domain R/P. Then there is a necessarily noninvertible ideal C of R such that $A = C/P$ and $C \supsetneq P$. Factor $C = HQ_1Q_2\cdots Q_m$ with H invertible and the Q_j pairwise comaximal primes. Taking images modulo P produces a corresponding factorization of $A = C/P$. Hence R/P has pseudo-Dedekind factorization. □

Theorem 5.1.4. *Let R be an integral domain with pseudo-Dedekind factorization.*

(1) Each nonzero finitely generated ideal of R has a nontrivial inverse. Hence each maximal ideal is a t-ideal.
(2) If M is a branched maximal ideal of R, then M is the radical of an invertible ideal.
(3) If $P \subsetneq P'$ are nonzero primes of R, then there is an invertible ideal C such that $P \subsetneq C \subsetneq P'$. Moreover, for each $g \in P' \setminus P$, the ideal $J := gR + P$ is invertible.
(4) If P is a nonzero nonmaximal prime of R, then $P = P(R:P)$ is antesharp and divisorial.
(5) R has the radical trace property.
(6) Each nonzero branched prime of R is sharp.
(7) R has strong factorization.

Proof. Let I be a nonzero finitely generated ideal of R. If I is invertible, it has a nontrivial inverse. If I is not invertible, then I^2 is not invertible and is properly contained in I. Hence there is an invertible ideal $A \neq R$ and comaximal primes $P_1, P_2, \ldots, P_n$ such that $I^2 = AP_1P_2\cdots P_n$. Thus $((R:I):I) = (R:I^2) \supseteq (R:A) \supsetneq R$, and therefore $(R:I) \supsetneq R$. This proves (1).

For (2), let M be a branched maximal ideal of R, and let $Q \subsetneq M$ be an M-primary ideal. Certainly, there is nothing to prove if Q is invertible, but if it is not, it factors as $Q = BM$ with $B(\neq R)$ invertible, necessarily with $\sqrt{B} = M$. Since M is maximal, B is M-primary.

Let $P \subsetneq P'$ be a pair of nonzero primes and let $J := b^2R + P$ where $b \in P' \backslash P$. Let $N \subseteq P'$ be a minimal prime of J and let $Q := JR_N \bigcap R$. Then Q is a proper N-primary ideal that (properly) contains P. If Q is not invertible, $Q = CN_1N_2 \cdots N_n$ for some invertible ideal C and pairwise comaximal primes $N_1, N_2, \ldots, N_n$. If C is not contained in N, then $QR_N = NR_N$, a contradiction. Hence $P \subsetneq Q \subseteq C \subseteq N \subseteq P'$. Thus by Proposition 2.3.20, each (prime) ideal that properly contains P blows up in $(P : P)$ and for $g \in R\backslash P$, the ideal $gR + P$ is invertible. It follows easily that P cannot be invertible in R. Hence it cannot be the case that PP^{-1} blows up in $(P : P)$, and therefore $PP^{-1} = P$. Also, P is divisorial by Corollary 2.3.21(1). Thus (3) and (4) hold.

Let I be a nonzero noninvertible ideal and let $I = BP_1P_2 \cdots P_n$ be a factorization with B invertible and the P_i pairwise comaximal primes. We may further assume no P_i is invertible, and thus by (4), $P_iP_i^{-1} = P_i$. Since the P_i are pairwise comaximal, $Q := P_1P_2 \cdots P_n = P_1 \bigcap P_2 \bigcap \cdots \bigcap P_n$ with $QQ^{-1} = Q$. Since B is invertible, $I^{-1} = B^{-1}Q^{-1}$, and hence $II^{-1} = QQ^{-1} = Q$ is a radical ideal. Therefore R has the radical trace property. Statement (6) is a consequence of (5) and Theorem 2.4.10.

Finally, to see that R has strong factorization we continue with the ideal I and its factorization as $BP_1P_2 \cdots P_n$. As above, we have $Q = P_1 \bigcap P_2 \bigcap \cdots \bigcap P_n = P_1P_2 \cdots P_n$, and therefore, again applying comaximality of the P_i (and the $(P_i)^v$), we have $Q^v = (P_1P_2 \cdots P_n)^v = (P_1^v \bigcap P_2^v \bigcap \cdots \bigcap P_n^v)^v = P_1^v \bigcap P_2^v \bigcap \cdots P_n^v = P_1^vP_2^v \cdots P_n^v$. Each nonmaximal P_i is divisorial, and if some P_j is maximal, then either $P_j = P_j^v$ or $P_j^v = R$. Thus Q^v is simply the product of those P_i that are divisorial with $Q^v = R$ if no P_i is divisorial. Since B is invertible, $I^v = (BQ)^v = BQ^v$. Let $\{M_1, M_2, \ldots, M_k\}$ be the (possibly empty) set of maximal P_i that are not divisorial and let $\{Q_1, Q_2, \ldots, Q_j\}$ be the (possibly empty) set of divisorial P_i. Then $Q^v = Q_1Q_2 \cdots Q_j$ ($= R$, if there are no divisorial P_i). We now have $I = BP_1P_2 \cdots P_n = BQ_1Q_2 \cdots Q_jM_1M_2 \cdots M_k = BQ^vM_1M_2 \cdots M_k = I^vM_1M_2 \cdots M_k$. Hence R has strong factorization. □

Corollary 5.1.5. *Let R be a domain with pseudo-Dedekind factorization.*

(1) If P is a nonzero nonmaximal prime of R, then $(R : P) = (P : P)$ and $J(P : P) = (P : P)$ for each ideal $J \supsetneq P$.
(2) If N is a finitely generated nonzero prime of R, then N is maximal.

Proof. The statement in (1) follows from the fact that P is antesharp. For (2), if N is a finitely generated nonzero prime of R, then $(N : N)$ is contained in the integral closure of R. Thus each maximal ideal of R that contains N survives in $(N : N)$. It follows that N is a maximal ideal of R. □

Theorem 5.1.6. *Let R be a local integral domain with pseudo-Dedekind factorization. If some nonzero finitely generated ideal J of R is not invertible, then R is not integrally closed, the maximal ideal M of R is two-generated and $J = bM$ for some $b \in R$.*

Proof. Let M be the maximal ideal of R. If some nonzero finitely generated ideal of R is not invertible, then there is two-generated ideal that is not invertible [34, Theorem 22.1]. Assume $I := (a, b)$ is a nonzero finitely generated ideal that is not invertible. Then $I = BP$ for some invertible ideal B and nonzero prime P. As R is local, $B = tR$ for some nonzero element t. It follows that $P = (a/t, b/t)$ is two-generated (and not invertible). By Corollary 5.1.5(2), $P = M$. For an arbitrary finitely generated noninvertible ideal $J \neq (0)$, we must have $J = qM$ for some q and therefore J is two-generated as well. Also, $(J : J) = (M : M)$. Since M is finitely generated and not invertible, $R \subsetneq (R : M) = (M : M)$ by Theorem 5.1.4(1). As the integral closure of R contains $(M : M)$, R is not integrally closed. □

Corollary 5.1.7. *Let R be an integrally closed domain. If R has pseudo-Dedekind factorization, then it is an h-local Prüfer domain.*

Proof. Assume R has pseudo-Dedekind factorization, and let M be a maximal ideal of R. Then R_M is integrally closed and has pseudo-Dedekind factorization. Moreover, each nonzero finitely generated ideal of R_M is invertible by Theorem 5.1.6. Hence R_M is a valuation domain. Therefore R is a Prüfer domain. By Theorem 5.1.4(7), R has strong factorization and must therefore be h-local (Corollary 4.4.10). □

Theorem 5.1.8. *Let R be an integrally closed domain. Then R has pseudo-Dedekind factorization if and only if R is an h-local Prüfer domain such that R_M has pseudo-Dedekind factorization for each maximal ideal M.*

Proof. Assume R has pseudo-Dedekind factorization. By Corollary 5.1.7 and Theorem 5.1.4, R is an h-local Prüfer domain such that R_M has pseudo-Dedekind factorization for each maximal ideal M.

For the converse, assume R is an h-local Prüfer domain such that R_M has pseudo-Dedekind factorization for each maximal ideal M. Let I be a noninvertible ideal of R. Since R is h-local, I is contained in only finitely many maximal ideals, say, $M_1, M_2, \ldots, M_n$, and it has the same number of minimal primes $Q_1, Q_2, \ldots, Q_n$ with $Q_i \subseteq M_i$. Also, $J := II^{-1}$ is a radical ideal of R (Theorem 2.4.12), necessarily with $J = P_1 P_2 \cdots P_m$ (with $m \leq n$) where the P_i are the (pairwise comaximal) minimal primes of II^{-1}. Each P_i is contained in a unique M_j. Since R is h-local, $I^{-1}R_M = (IR_M)^{-1}$ for each maximal ideal M (Proposition 2.1.8). Also, Theorem 5.1.6 yields that for each M_i, IR_{M_i} is either principal or a principal multiple of a nonprincipal prime since R_{M_i} has pseudo-Dedekind factorization. The former occurs when II^{-1} is not contained in M_i, and the latter when II^{-1} is contained in M_i. Split the set $\{M_1, M_2, \ldots, M_n\}$ into two sets, the set $\{M_1', M_2', \ldots, M_m'\}$ of those M_i that contain II^{-1} (and the corresponding P_i) and the (possibly empty) set $\{N_1, N_2, \ldots, N_t\}$ of those M_j that do not contain II^{-1}. In $R_{M_i'}$, $IR_{M_i'} = b_i P_i R_{M_i'}$ for some $b_i \in R$. If $b_i \in M_i'$, then it has a unique (nonzero) minimal prime contained in M_i' that is contained in no other maximal ideal. Hence there is a finitely generated ideal $B_i \subseteq M_i'$ such that $B_i R_{M_i'} = b_i R_{M_i'}$ where M_i' is the only maximal ideal that contains B_i. In the event b_i is not in M_i, set $B_i := R$. For each N_j, IR_{N_j} is principal and, as with the case $b_i \in M_i'$, there is a finitely generated ideal $C_j \subseteq N_j$, where $C_j R_{N_j} = IR_{N_j}$, with N_j the only maximal ideal containing C_j. The ideal $A := \prod_i B_i \prod_j C_j$ is invertible, and checking locally shows that $I = AP_1P_2 \cdots P_m$. Hence R has pseudo-Dedekind factorization. □

5.2 Local Domains with Pseudo-Dedekind Factorization

In 1976, P. Eakin and A. Sathaye [17] introduced the notion of a *prestable ideal* as an ideal I such that for each prime P of a ring R (not necessarily a domain), there is a positive integer n such that $I^{2n}R_P = dI^nR_P$ for some $d \in I^n$. (Recall that a *stable ideal* of a ring R (in the sense of [56]) is an ideal I such that for each prime ideal P of R, $I^2R_P = dIR_P$ for some $d \in R$; therefore a prestable ideal is an ideal for which locally some power is stable, with the power allowed to vary from prime to prime.) It is clear that I is prestable if and only if IR_M is prestable for each maximal ideal M of R. If I is finitely generated and contains an element that is not a zero divisor, then it is prestable if and only if for each maximal ideal M, there is a positive integer n and an element $b \in I^n$, such that $I^nR_M = b(I^nR_M : I^nR_M)$ (see the proof of [17, Corollary 1]). We note that if $I^nR_M = b(I^nR_M : I^nR_M)$, then $I^{2n}R_M = bI^nR_M$. Moreover, if $d \in I^n$ is such that $I^{2n}R_M = dI^nR_M$, then $d/1$ is not a zero divisor of R_M. Hence $(1/d)I^{2n}R_M = I^nR_M$, which implies $I^nR_M \subseteq d(I^nR_M : I^nR_M) \subseteq I^nR_M$. While [17, Theorem 2] is stated only for semilocal domains, it holds for all integral domains—the integral closure of a domain R is a Prüfer domain if and only if each nonzero finitely generated ideal of R is prestable.

As we saw in Theorem 5.1.6 above, if R is a local domain with pseudo-Dedekind factorization that is not integrally closed, then each finitely generated noninvertible ideal is both two generated and a principal multiple of the (two-generated) maximal ideal.

Lemma 5.2.1. *Let R be an integral domain. If each two-generated ideal I of R contains an element d such that $dI = I^2$, then the integral closure of R is a Prüfer domain.*

Proof. Recall that if each two-generated ideal of a domain is invertible, then the domain is Prüfer [34, Theorem 22.1]. Denote by R' the integral closure of R, let $J := fR' + gR'$ be a two-generated ideal of R', and let $I := pfR + pgR$, where p is a nonzero element of R that multiplies both f and g into R. By assumption, there is an element $d \in I$ such that $I^2 = dI = dpfR + dpgR$. Then we have $(1/d)I \cdot I = I$. Thus $(1/d)I \subseteq (I : I)$. Since I is finitely generated, $(I : I) \subseteq R'$. It follows that $1/d \in (R' : I)$ with $d \in I$, and we have that $pJ = IR' = dR'$ is an invertible ideal of R'. Hence J is invertible as an ideal of R'. Therefore R' is a Prüfer domain. □

Recall from Proposition 2.4.1 that if I is a nonzero noninvertible ideal of a valuation domain V, then II^{-1} is a prime ideal of V. The next two lemmas provide

is two-generated. It follows that $(J : J) = (M : M)$ for each such J. For an invertible ideal I, $(I : I) = R$. Hence $R' = (M : M)$. Also R' is Prüfer domain by Lemma 5.2.1.

We next show that R' is local with maximal ideal M. For each unit u of R', $uM = M$, but for a nonzero nonunit $t \in R'$, tM is properly contained M. Since M is not invertible as an ideal of R, neither is tM. Hence $tM = sM$ for some $s \in M$. It follows that $(t/s)M = M$ and we have that $t/s = w$, a unit of R'. Thus $t = ws \in M$ and we have that M is the maximal ideal of R'. As R' is a Prüfer domain, it is a valuation domain and R is PVD, and R and R' have the same prime ideals. If I is not invertible in R', then it certainly is not invertible in R, and therefore $I = bP$ for some $b \in R$ and prime P. Clearly, the same factorization is valid in R'. Hence R' has pseudo-Dedekind factorization.

Let $a, b \in R$ generate M as an ideal of R. Since $M^2 = pM$ for some $p \in M$, we have $(1/p)M \cdot M = M$. Hence $1/p \in (R' : M)$ and we have $aR + bR = M = pR'$ with $a/p, b/p \in R'$. It follows that $R' = (a/p)R + (b/p)R$. Hence $[R'/M : R/M] = 2$.

For the converse, assume that R' is a valuation domain with principal maximal ideal $M = pR'$ such that R' has pseudo-Dedekind factorization and $[R'/M : R/M] = 2$. Then there is a unit $q \in R'$ such that $R' = R+qR$. Since M is the common maximal ideal of R and R', R is a PVD. It follows that $M = pR+pqR$ is two-generated in R. Also, there are no R-modules strictly between R and R'.

Each nonzero ideal I of R is of one of the following three types: (i) $I = IR'$ is not invertible as an ideal of R', (ii) $I = IR'$ is invertible as an ideal of R', and (iii) $I \subsetneq IR'$ (equivalently, I is not an ideal of R').

As R and R' have the same prime ideals, there is nothing to prove if I is a prime ideal of R. Hence we assume that I is not prime.

Since R' has pseudo-Dedekind factorization, if $I = IR'$ is not invertible as an ideal R', then $I = bP$ for some $b \in R'$ and prime P. Since we have assumed I is not prime, $b \in M$, and the factorization $I = bP$ is of the desired type in R as well.

If $I = IR'$ is invertible as an ideal of R', then $I = bR'$ for some $b \in I$. Since I is not prime, it is properly contained in $M = pR'$. Hence $b/p \in M$ and we have $I = p(b/p)R' = (b/p)M$ with $b/p \in R$.

The remaining case is when I is properly contained in IR'. Since $R' = R + qR$, it must be that qI is not contained in I. Let $f, g \in I$ with $g/f \in M$. Then $qg/f \in M$, and therefore $qg = f(qg/f) \in I$. Thus if $f \in I$ is such that $qf \notin I$, then we have $IR' = fR'$ and $fR \subseteq I \subsetneq fR' = fR + fqR$ since $R' = R + qR$. Multiplying by $1/f$ yields $R \subseteq (1/f)I \subsetneq R'$. As there are no R-modules properly between R and R', we have $(1/f)I = R$ and therefore $I = fR$ is a principal ideal of R. It follows that R has pseudo-Dedekind factorization. □

Corollary 5.2.5. *Let V be a strongly discrete valuation domain with (principal) maximal ideal M such that $[V/M : F] = 2$ for some subfield $F \subsetneq V/M$. Then the pseudo-valuation domain R that results from taking the pullback of F over M has pseudo-Dedekind factorization.*

more detailed information about certain types of nonzero noninvertible i a valuation domain. The first was used in [15] to characterize which v domains have the property that each nonzero noninvertible ideal I can be as a principal multiple of a P-primary ideal Q where $P = II^{-1}$.

Lemma 5.2.2. (Cahen–Lucas [15, Corollary 5]) *Let I be a nonzero nonin ideal of a valuation domain V, and let $P := II^{-1}$. Then $I = bQ$ for some and P-primary ideal Q if and only if there is an element $t \in I^{-1}$ such that t for each prime $P' \subsetneq P$.*

Proof. Obviously, if $I = bQ$ for some $b \in V$ and P-primary ideal Q, ther I^{-1} with $b^{-1}I = Q$. Since V is a valuation domain, Q properly contai prime P' that is properly contained in P.

For the converse, assume there is an element $t \in I^{-1}$ such that $tI \supsetneq P'$ f prime $P' \subsetneq P$. Since $II^{-1} = P$, $I \subseteq P$ and $\sqrt{tI} = P$. Let $c, d \in V$ be su $cd \in tI$ with $c \in V \backslash P$. Then $d \in P$. If d is not in tI, then we have $dV \supsetneq$ therefore $(1/d) \in (V : tI)$. It follows that $c = cd/d \in tI(tI)^{-1} = II^{-1}$ a contradiction. Hence $d \in tI$, and therefore $Q := tI$ is P-primary. If $t \in$ either $tV \supsetneq P$ or $\sqrt{tV} = P$; in either case, $\sqrt{I} = P$, and so the argume presented shows that I is P-primary. If t is not in V, then $1/t \in V$ and w $I = (1/t)Q$.

We can say more when the prime $P = II^{-1}$ is such that PV_P is principal.

Lemma 5.2.3. *Let I be a nonzero noninvertible ideal of a valuation dom and let $P := II^{-1}$. If PV_P is principal, then there is an element $b \in V$ suc $I = bP$.*

Proof. Assume $PV_P = qV_P$ with $q \in P$. Then $P = \sqrt{qV}$. There is noth prove if $I = P$, so we may assume $I \subsetneq P$. Since the finitely generated ideals are all principal, there are elements $w \in I$ and $y \in I^{-1}$ such that $qV \subseteq wy$ each prime $P' \subsetneq P$, we have $P' \subsetneq qV \subseteq yI$. From the proof of Lemma 5.2. is P-primary, and so $yI = P$ and $y \notin V$. Thus $I = (1/y)P$, necessarily wit in V.

Recall from [43] that a local domain (R, M) is a *pseudo-valuation do* (or *PVD*) if $(M : M)$ is a valuation domain (called the canonical valu overring of R) with maximal ideal M. A PVD R and its overring $(M : M)$ the same set of prime ideals [43, Theorem 2.7]. Also, if $R \subsetneq (M : M)$, $(M : M) = (R : M)$.

Theorem 5.2.4. *Let R be a local domain that is not integrally closed and le be its maximal ideal and R' its integral closure. Then R has pseudo-Dede factorization if and only if R' is a valuation domain with principal maximal i M, R' has pseudo-Dedekind factorization, and $[R'/M : R/M] = 2$.*

Proof. Assume R has pseudo-Dedekind factorization. Then by Theorem 5.1.6, two-generated and not invertible as an ideal of R. Moreover, each finitely gener noninvertible ideal J of R has the form $J = bM$ for some $b \in R$ and s

Proof. Since V is a strongly discrete valuation domain, if I is a nonzero noninvertible ideal of V, then $I(V : I) = P$ is prime and there is an element $b \in V$ such that $I = bP$ (Proposition 2.4.1 and Lemma 5.2.3). Hence V has pseudo-Dedekind factorization. That R has pseudo-Dedekind factorization follows from Theorem 5.2.4. □

For a simple example, the rings $R := \mathbb{Q} + X\mathbb{Q}[\sqrt{2}][[X]] \subsetneq R' = \mathbb{Q}[\sqrt{2}][[X]]$ are local with pseudo-Dedekind factorization.

Our next goal is to characterize which valuation domains have pseudo-Dedekind factorization. We again make use of the fact that a valuation domain V has the trace property (Proposition 2.4.1). In fact, if P is prime and Q is a noninvertible P-primary ideal, then it is not difficult to see that $QQ^{-1} = P$. It follows that if $J = tQ$ for some $t \in V$, then $JJ^{-1} = P$ also.

Lemma 5.2.6. *Let V be a valuation domain with pseudo-Dedekind factorization. If P is a nonmaximal prime of V, then PV_P is principal in V_P.*

Proof. Let P be a (nonzero) nonmaximal prime of V, and let $I := qV_P$, where q is a nonzero element of P. Then $I(V : I) = qV_P(V : qV_P) = V_P(V : V_P) = PV_P = P$. Thus I is a noninvertible ideal of V. Since V has pseudo-Dedekind factorization, we have $I = rP$ for some $r \in V$. Hence $qV_P = I = rP = rPV_P$, and PV_P is principal. □

The next theorem characterizes which rank one valuation domains have pseudo-Dedekind factorization.

Theorem 5.2.7. *Let V be a rank one valuation domain with corresponding value group G. Then V has pseudo-Dedekind factorization if and only if G is isomorphic to either $\mathbb{Z}$ or $\mathbb{R}$.*

Proof. Obviously, a discrete rank one valuation domain has pseudo-Dedekind factorization. Thus we need consider only the case where the maximal ideal M of V is idempotent. We may then consider the value group G to be a dense subgroup of $\mathbb{R}$.

First, suppose V has pseudo-Dedekind factorization, and let α be a positive real number. Also, let $A := \{a \in G^+ \mid \alpha < a\} \bigcup \{\infty\}$. Since G is dense in $\mathbb{R}$, $\alpha = \inf\{a \in A\}$. Now, let v denote the corresponding valuation, and set $I := v^{-1}(A)$. Then I is not a principal ideal of V, so by pseudo-Dedekind factorization, $I = bM$ for some element $b \in M$. For $c := bm, m \in M$, we have $v(b) < v(c) = v(b) + v(m)$. As m may be chosen with arbitrarily small positive value, it must be that $v(b) = \alpha$. Hence $G = \mathbb{R}$.

For the converse, assume $G = \mathbb{R}$ and let J be a nonzero noninvertible ideal of V. Then (continuing to denote the valuation by v) the set $C := v(J)$ contains no minimum value but does have an infimum β. Since $G = \mathbb{R}$, there is an element $b \in V$ with $v(b) = \beta$. It follows that $J = bM$, and therefore V has pseudo-Dedekind factorization. □

For a prime ideal P of a valuation domain V, the intersection of the P-primary ideals is a prime ideal P_0 that contains every prime that is properly contained in P. Obviously, $P_0 \subsetneq P$ if and only if P is branched. We shall continue to use this notation for the rest of this chapter.

Corollary 5.2.8. *Let V be a valuation domain with pseudo-Dedekind factorization. If the maximal ideal M is branched but not principal, then V/M_0 is a rank one valuation domain whose value group is (isomorphic to) $\mathbb{R}$.*

Proof. Since M is branched, $M_0 \subsetneq M$ and V/M_0 is a rank one valuation domain. By Lemma 5.1.3, V/M_0 has pseudo-Dedekind factorization, and it is clear that M/M_0 is branched but not principal. Thus the value group of V/M_0 is isomorphic to $\mathbb{R}$ by Theorem 5.2.7. □

Theorem 5.2.9. *Let V be a valuation domain whose maximal ideal M is branched. Then V has pseudo-Dedekind factorization if and only if PV_P is principal for each nonmaximal prime P and either M is principal or the value group of V/M_0 is (isomorphic to) $\mathbb{R}$.*

Proof. If V has pseudo-Dedekind factorization, then PV_P is principal for each nonmaximal prime P (Lemma 5.2.6). Also, by Corollary 5.2.8, either M is principal or the value group of V/M_0 is $\mathbb{R}$.

For the converse, let I be an nonzero noninvertible ideal of V, and set $Q := II^{-1}$. If QV_Q is principal, then $I = tQ$ for some $t \in V$ by Lemma 5.2.3. So, the only case we need consider is when $Q = M$ and PV_P is principal for each nonmaximal prime P but V/M_0 has value group isomorphic to $\mathbb{R}$. Since M is branched, it is the radical of a principal ideal sV. Thus there are elements $d \in I$ and $y \in I^{-1}$ such that $dyV \supseteq sV \supsetneq M_0$. It follows that yI is an M-primary ideal of V. As noted above, $(yI)(yI)^{-1} = II^{-1} = M$. Hence we may assume I is an M-primary ideal, in which case, $I \supsetneq M_0$. Moreover, $I^{-1}/M_0 = (I/M_0)^{-1}$. Thus $(I/M_0)(I/M_0)^{-1} = II^{-1}/M_0 = M/M_0$. By Theorem 5.2.7, $I/M_0 = bM/M_0$ for some $b \in V\backslash M_0$ and therefore $I = bM$. Hence V has pseudo-Dedekind factorization. □

We say that a valuation domain V is *principally complete* if whenever there are two families of nonzero elements $\{b_\alpha\}_{\alpha\in\mathscr{A}}$ and $\{c_\alpha\}_{\alpha\in\mathscr{A}}$ and a corresponding family of primes $\{P_\alpha\}_{\alpha\in\mathscr{A}}$ with $\mathscr{A}$ totally ordered such that for all $\alpha < \beta$ in $\mathscr{A}$:

(i) $b_\alpha \in P_\alpha$,
(ii) $b_\alpha V \subseteq b_\beta V \subseteq c_\beta V \subseteq c_\alpha V$,
(iii) $b_\alpha/c_\alpha \in V\backslash P_\alpha$, and
(iv) $P_\alpha \subseteq P_\beta$ with $P := \bigcup P_\alpha$ an unbranched prime,

then there is an element $c \in V$ such that $b_\alpha V \subseteq cV \subseteq c_\alpha V$ for all $\alpha \in \mathscr{A}$. Note that when such an element c exists, then $(b_\alpha/c_\alpha)V \subseteq (b_\alpha/c)V \subseteq V$ for all α. Thus by (iii), $b_\alpha/c \in V\backslash P_\alpha$ for all α.

Theorem 5.2.10. *The following statements are equivalent for a valuation domain V with unbranched maximal ideal M.*

(i) *V has pseudo-Dedekind factorization.*
(ii) *PV_P is principal in V_P for each nonmaximal prime P of V and, for each ideal I with $II^{-1} = M$, there is an element $t \in I^{-1}$ such $tI \supsetneq P$ for each prime $P \subsetneq M$.*
(iii) *V is principally complete and PV_P is principal in V_P for each nonmaximal prime P of V.*

Proof. By Lemma 5.2.6, if V has pseudo-Dedekind factorization, then PV_P is principal for each nonmaximal prime P. If I is such that $II^{-1} = M$, then I must factor as $I = pM$ for some element $p \in V$. It is clear that $1/p$ is in I^{-1} with $(1/p)I = M \supsetneq P$ for each nonmaximal prime P. Therefore (i) implies (ii).

We next show that (ii) implies both (i) and (iii). Assume (ii) holds and let I be an nonzero noninvertible ideal with the prime ideal $P := II^{-1}$ properly contained in M. Then P is branched and PV_P is principal. Thus $I = bP$ for some $b \in V$ by Lemma 5.2.3. If J is a nonzero noninvertible ideal with $JJ^{-1} = M$ and $J \subsetneq M$, we have an element $t \in J^{-1}$ such that $tJ \supsetneq Q$ for each prime $Q \subsetneq M$. Since M is unbranched, it is the union of the nonmaximal primes. Hence $tJ = M$, and therefore $J = (1/t)M$ with $1/t \in V$. Thus V has pseudo-Dedekind factorization.

To see that V is principally complete, let $B := \{b_\alpha\}_{\alpha\in\mathscr{A}}$, $C := \{c_\alpha\}_{\alpha\in\mathscr{A}}$, and $\{P_\alpha\}_{\alpha\in\mathscr{A}}$ be as in the definition above. Let I be the ideal generated by the set B. A consequence of condition (ii) of the definition is that $b_\gamma/c_\alpha \in V$ for all $\alpha, \gamma \in \mathscr{A}$. Hence $c_\alpha^{-1} \in I^{-1}$ for each α. It follows that II^{-1} properly contains each P_α and thus $II^{-1} \supseteq P = \bigcup P_\alpha$. Since M is the only unbranched prime of V, we either have $II^{-1} = V$ or $II^{-1} = M$.

If $II^{-1} = V$, then I is principal. We then have $I = b_\gamma V$ for some $b_\gamma \in B$. Obviously, $b_\alpha V \subseteq b_\gamma V \subseteq c_\alpha V$ for each α, and we may take $c := b_\gamma$ to satisfy the requirements of the definition in this case.

If $II^{-1} = M$, then $I = cM$ for some $c \in V$ since V has pseudo-Dedekind factorization. In this case, $1/c \in I^{-1}$ and $b_\alpha V \subsetneq cV$ for each α. Since I is not invertible, $(1/c_\alpha)I \subseteq M$, and therefore $cM = I \subseteq c_\alpha M$. As $M^{-1} = V$, we have $cV \subseteq c_\alpha V$ for each α. Thus V is principally complete.

All that remains is to show that (iii) implies (ii) (or (i)). For this, all we need prove is that if I is an ideal such that $II^{-1} = M$, then $I = cM$ for some $c \in V$. Let P be the minimal prime of I. Since M is unbranched, if $P = M$, then we also have $I = M$. Thus we may assume $P \subsetneq M$. Let $\{Q_\gamma\}$ denote the set of nonmaximal primes that properly contain P. Then $\bigcup Q_\gamma = M$. Since $II^{-1} = M$, IV_{Q_γ} is invertible and therefore principal in V_{Q_γ}.

Each Q_γ has an immediate predecessor, but there may be some with no immediate successor. So the first step is to reduce the full set $\{Q_\gamma\}$ to the set of those primes Q_γ that do have an immediate successor. Denote this set/family as $\{P_\alpha\}_{\alpha\in\mathscr{A}}$. For each P_α, denote its immediate successor as $P_{\alpha+1}$. For each P_α, the ideal IV_{P_α} is principal in V_{P_α} as is $P_\alpha V_{P_\alpha}$.

Let $p_\alpha \in P_\alpha$ be such that $p_\alpha V_{P_\alpha} = P_\alpha V_{P_\alpha}$. Next let $b'_\alpha \in I$ be such that $b'_\alpha V_{P_{\alpha+1}} = IV_{P_{\alpha+1}}$. Set $b_\alpha := b'_\alpha p_{\alpha+1}$. Since $p_{\alpha+1}$ is a unit in V_{P_α} and $V_{P_{\alpha+1}} \subsetneq V_{P_\alpha}$,

$b_\alpha V_{P_\alpha} = b'_\alpha V_{P_\alpha} = IV_{P_\alpha}$, but $b_\alpha V_{P_{\alpha+1}} \subsetneq IV_{P_{\alpha+1}}$. Also, for $\alpha < \beta$, $b_\alpha V_{P_\beta} \subsetneq b_\beta V_{P_\beta}$ and $b_\beta V_{P_\alpha} = IV_{P_\alpha}$ since $V_{P_\alpha} \supsetneq V_{P_\beta}$. It follows that $b_\alpha V \subsetneq b_\beta V$.

Next, let $c_\alpha := b'_\alpha / p_{\alpha+1}$. In $V_{P_{\alpha+1}}$, $c_\alpha V_{P_{\alpha+1}}$ properly contains $b'_\alpha V_{P_{\alpha+1}} = IV_{P_{\alpha+1}}$. Hence c_α is not in I. On the other hand, for $\alpha < \beta$, $c_\beta V_{P_\alpha} = b_\beta V_{P_\alpha} = b_\alpha V_{P_\alpha} = IV_{P_\alpha} \subsetneq c_\alpha V_{P_\alpha}$. Thus $b_\alpha V \subsetneq b_\beta V \subsetneq c_\beta V \subsetneq c_\alpha V$ for all $\alpha < \beta$. Also $b_\alpha / c_\alpha = p^2_{\alpha+1} \in V \backslash P_\alpha$.

Thus the families $B = \{b_\alpha\}_{\alpha \in \mathscr{A}}$, $C = \{c_\alpha\}$ and $\{P_\alpha\}_{\alpha \in \mathscr{A}}$ satisfy conditions (i)–(iv). Hence there is an element $c \in V$ such that $b_\alpha V \subsetneq cV \subsetneq c_\alpha V$ and $b_\alpha / c \in V \backslash P_\alpha$ for each α. Since $M = \bigcup P_\alpha$, it must be that $c^{-1} I = M$, and therefore $I = cM$. □

In the next example we construct a valuation domain which has both an unbranched maximal ideal and pseudo-Dedekind factorization. For such a valuation domain V, if $R \subsetneq V$ is pseudo-valuation domain with the same maximal ideal, then R does not have pseudo-Dedekind factorization (Theorem 5.2.4).

Example 5.2.11. Let V be a valuation domain with corresponding value group $G := \prod_{n=1}^{\infty} G_n$ under complete lexicographic order with $G_n := \mathbb{Z}$ for each n. By [15, Theorem 7], each nonzero noninvertible ideal I factors as a principal multiple of a primary ideal. We also have that each nonmaximal prime ideal P of V is such that PV_P is principal in V_P. It then follows from Lemma 5.2.3 that V has pseudo-Dedekind factorization. However, the maximal ideal M is not principal. Indeed, $M = \bigcup_n P_n$, where the P_n are the nonzero nonmaximal primes of V.

Lemma 5.2.12. *Let $P \subseteq N$ be a pair of nonzero primes of a domain R. If $bR \supsetneq P$ for each $b \in N \backslash P$, then $P = \bigcap\{bR \mid b \in N \backslash P\}$.*

Proof. Clearly, if $bR \supsetneq P$ for each $b \in N \backslash P$, then $P \subseteq \bigcap\{bR \mid b \in N \backslash P\}$. To achieve equality, simply note that if $t \in N \backslash P$, then $t^2 R \subsetneq tR$. Thus if $s \in \bigcap\{bR \mid b \in N \backslash P\}$, then $s \in P$. □

Theorem 5.2.13. *Let R be an integral domain with pseudo-Dedekind factorization. Then the integral closure R' of R is an h-local Prüfer domain with pseudo-Dedekind factorization. Moreover, each maximal ideal M of R is contained in a unique maximal ideal $M' = MR'$ of R', R_M is a PVD with pseudo-Dedekind factorization, and the canonical valuation overring of the PVD R_M is $R'_{M'} = R'_M$ with common maximal ideal $MR_M = M'R'_{M'}$.*

Proof. Let M be a maximal ideal of R. Then R_M has pseudo-Dedekind factorization (Lemma 5.1.3). Hence by Theorem 5.2.4, R_M is a pseudo-valuation domain whose integral closure, R'_M, is a valuation domain with maximal ideal MR'_M. As each maximal ideal of R' that lies over M extends to a maximal ideal of R'_M, each maximal ideal of R is contained in a unique maximal ideal of R'. In particular, for $M \in \operatorname{Max}(R)$, $M' = MR'_M \bigcap R' = MR'$ is the unique maximal ideal of R' that lies over M. Since each maximal ideal of R' lies over a maximal ideal of R, R' is a Prüfer domain.

We next show that each nonzero nonmaximal prime of R is contained in a unique maximal ideal. By way of contradiction, suppose there is a nonzero nonmaximal prime P of R that is contained in distinct maximal ideals M and N. For $S := R\backslash(M \bigcup N)$, R_S has pseudo-Dedekind factorization and exactly two maximal ideals MR_S and NR_S both containing PR_S. Hence we may reduce to the case that R has exactly two maximal ideals. Since both R_M and R_N are pseudo-valuation domains, we may further assume that P is maximal with respect to being contained in both M and N. Thus P compares with each prime of R. Also, for each $b \in R\backslash P$, $(1/b)PR_M = PR_M$ and $(1/b)PR_N = PR_N$. Hence $bR \supsetneq P$.

Since R_M is a pseudo-valuation domain, R_P is a valuation domain with PR_P principal. Thus $P = \sqrt{dR}$ for some element $d \in P$ with $PR_P = dR_P$. Consider the ideal $I := dR_M \bigcap R$. For each $c \in N\backslash M$, we clearly have $d/c \in dR_M$, but also $d/c \in P$ from above. Thus since $d/c \in I\backslash dR$, $I \supsetneq dR$. Also, for $f \in M\backslash P$, while d/f is in P, it is not in I. Thus $dR \subsetneq I \subsetneq P$. The ideal I is generated by the set $\{d/c^n \mid c \in N\backslash M\}$ and certainly no element in this set can generate IR_N. Hence IR_N is not principal, and therefore I is not invertible as an ideal of R.

Potentially, there are four ways I might factor as the product of a principal ideal hR and pairwise comaximal primes (or a single prime):

(1) $I = hP$,
(2) $I = hQ$ where $P \subsetneq Q \subseteq M$,
(3) $I = hQ'$ where $P \subsetneq Q' \subseteq N$, and/or
(4) $I = hQQ'$ where $P \subsetneq Q \subseteq M$ and $P \subsetneq Q' \subseteq N$.

In the first, $II^{-1} = P$; and in the other three, $II^{-1} \supsetneq P$. If $I = hP$ with $h \notin P$, then (checking locally) we have $I = P$, a contradiction. On the other hand, if $I = hP$ with $h \in P$, then we have $IR_P \subsetneq PR_P = dR_P = IR_P$, again a contradiction. To eliminate the other three factorizations consider an element $t \in I^{-1}$. Then $r := td \in R$. Thus for each $c \in N\backslash P$, we have $td/c = r/c \in R$, putting $r \in \bigcap\{cR \mid c \in N\backslash P\} = P$ by Lemma 5.2.12. Hence none of the factorizations are valid. Therefore it must be that each nonzero prime of R is contained in a unique maximal ideal (of R).

Let P' be a nonzero nonmaximal prime of R' and let $P := P' \bigcap R$. Since P is contained in a unique maximal ideal of R, which in turn is contained in a unique maximal ideal of R', P' must be contained in a unique maximal ideal of R'.

Let P' be a nonzero branched prime of R', and let $P := P' \bigcap R$. If P' is a maximal ideal of R', then R_P is a pseudo-valuation domain with maximal ideal $PR_P = P'R'_{P'}$. Thus P is a branched maximal ideal of R. By Theorem 5.1.4, such a maximal ideal is the radical of an invertible ideal. Hence P' is the radical of an invertible ideal of R'. If, instead, P' is not maximal, then $R_P = R'_{P'}$. Hence, again, P is branched and by Theorem 5.1.4(7), R_P does not contain $\Theta(P)$. As $\Theta(P')$ contains $\Theta(P)$ (and $R_P = R'_{P'}$), P' is a sharp prime of R'. Therefore R' has the radical trace property (Theorem 2.4.10). By Theorem 2.4.12, R' is an h-local Prüfer domain. Also, R' has pseudo-Dedekind factorization by Theorem 5.1.8. □

A domain R such that R_M is a pseudo-valuation domain for each maximal ideal M is said to be a *locally pseudo-valuation domain* [16]. Thus a necessary condition for R to have pseudo-Dedekind factorization is for it to be a locally pseudo-valuation domain.

Corollary 5.2.14. *Let R be an integral domain. Then R has pseudo-Dedekind factorization if and only if it is h-local and R_M has pseudo-Dedekind factorization for each maximal ideal M.*

Proof. By Theorem 5.1.8, there is nothing to prove if R is integrally closed, so we assume it is not.

Assume R has pseudo-Dedekind factorization. Then R_M has pseudo-Dedekind factorization for each maximal ideal. Also, from the proof of Theorem 5.2.13, each nonzero prime of R is contained in a unique maximal ideal (of both R and R'). Since each nonzero nonunit of R' is contained in only finitely many maximal ideals of R', the same occurs for each nonzero nonunit of R. Hence R is h-local.

For the converse, assume R is h-local and R_M has pseudo-Dedekind factorization for each maximal ideal M. Let I be a nonzero noninvertible ideal of R. Then I is contained in only finitely many maximal ideals, $M_1, M_2, \ldots, M_n$. Also, Proposition 2.1.8 assures that, for each maximal ideal M, $I^{-1}R_M = (IR_M)^{-1}$ (equal to R_M, except possibly for $M = M_i$ for some i). It follows that since I is not invertible, there is at least one M_i such that IR_{M_i} is not invertible. Renumber if necessary to have $\{M_1, M_2, \ldots, M_k\}$ be the set of those M_i such that IR_{M_i} is not invertible and $\{M_{k+1}, \ldots, M_n\}$ be the (possibly empty) set of M_i where IR_{M_i} is invertible. For each $M_i \in \{M_1, M_2, \ldots, M_k\}$, there is an invertible ideal A_i and prime $P_i \subseteq M_i$ such that $IR_{M_i} = A_i P_i R_{M_i}$ since R_{M_i} has pseudo-Dedekind factorization. Using finite character, we may further assume that M_i is the only maximal ideal that contains A_i. Since R is h-local, M_i is the only maximal ideal that contains P_i. For $M_j \in \{M_{k+1}, \ldots, M_n\}$, all we need is an invertible ideal $B_j \subseteq M_j$ with $IR_{M_j} = B_j R_{M_j}$ and with M_j the only maximal ideal containing B_j. Let $A := \prod_i A_i$ and $B := \prod_j B_j$ ($= R$, if $k = n$). Checking locally, we see that $IR_M = ABP_1P_2\cdots P_k R_M$ for each maximal ideal M, so that $I = ABP_1P_2\cdots P_k$. Therefore R has pseudo-Dedekind factorization. □

5.3 Strong Pseudo-Dedekind Factorization

For a nonzero ideal I of an integral domain R, a *special factorization* of I is a factorization $I = BP_1P_2\cdots P_n$ with B a finitely generated ideal of R (possibly with $B = R$) and $P_1, P_2, \ldots, P_n$ primes with $n \geq 1$. We then say that the domain R has *special factorization* if each nonzero ideal of R has a special factorization. Clearly, a ZPUI-domain, e.g., a domain with strong pseudo-Dedekind factorization, has special factorization.

It is not completely trivial that special factorization is preserved with regard to quotient ring formation. One must do some work to guarantee there is a factorization of a nonzero ideal of R_S that contains at least one prime of R_S.

Lemma 5.3.1. *Let R have special factorization. Then*

(1) R/P has special factorization for each nonmaximal prime P of R, and
(2) R_S has special factorization for each multiplicative set S of R.

Proof. Let P be a prime and let $I \supsetneq P$. Then there is a finitely generated ideal J and prime ideals $P_1, P_2, \cdots, P_n$ such that $I = JP_1P_2\cdots P_n$. Moding out by P, we obtain the factorization $I/P = (J/P)(P_1/P)(P_2/P)\cdots(P_n/P)$ in R/P.

Let S be a multiplicative set in R and I an ideal such that $I \bigcap S = \emptyset$. There is no problem if there is a factorization of I as $JP_1P_2\cdots P_n$ with J finitely generated and $P_i \bigcap S = \emptyset$ for at least one i. On the other hand, if each P_i has a nonempty intersection with S, then $J \bigcap S = \emptyset$, $IR_S = JR_S$ and $P_iR_S = R_S$ for each P_i. Let $H := IR_S \bigcap R$. Factor $H = BQ_1Q_2\cdots Q_m$ with B finitely generated and $Q_1, Q_2, \ldots, Q_m$ primes. Then $IR_S = HR_S = BQ_1Q_2\cdots Q_mR_S$. If each Q_i blows up in R_S, then $BR_S = HR_S$ which implies $B = H = BQ_1Q_2\cdots Q_m$. Since B is finitely generated, this is impossible (by Nakayama's Lemma). Hence at least one Q_j survives in R_S and IR_S has a special factorization in R_S. □

The first theorem makes use of Lemma 3.1.6 to show that at least some maximal ideals in a domain with special factorization have properties similar to those in a domain with property (α) (i.e., those domains where each primary ideal is a power of its radical, see Sect. 3.1).

Theorem 5.3.2. *Let R be a local domain such that the maximal ideal M is the radical of finitely generated ideal. If R has special factorization, then M is principal and $P := \bigcap_n M^n$ is a prime that contains each nonmaximal prime and is properly contained in M.*

Proof. By Lemma 3.1.6, it suffices to show that $\{M^k\}$ is the complete set of M-primary ideals. We first show that M is finitely generated.

Let I be a finitely generated ideal with $\sqrt{I} = M$. Then without loss of generality, we may assume $I \neq M$. Thus $I = JM^k$ for some finitely generated ideal J of R and positive integer k. If $M = M^2$, then $IM = JM^{k+1} = JM^k = I$ which gives a contradiction by Nakayama's Lemma. Thus $M \neq M^2$.

Next, let $r \in M \backslash M^2$. Then $Q := rR + M^2$ is M-primary and $Q = AM^j$ for some finitely generated ideal A and some positive integer j. Clearly, we must have $j = 1$ and $A = R$. Hence $M = rR + M^2$.

Consider the ideal $B := rR + I$. Then, clearly, M is the radical of B and M^2 does not contain B. By special factorization, $B = CM^j$ for some finitely generated ideal C and some positive integer j. Since B is not contained in M^2, it must be that $C = R$ and $j = 1$; i.e., $B = M$. Thus M is finitely generated.

Next, suppose Q_1 is a proper M-primary ideal. Then $Q_1 = A_1M^{k_1}$ for some integer $k_1 \geq 1$ and finitely generated ideal A_1. Since M is finitely generated, there is a smallest integer j such that $Q_1 \supseteq M^j$. Thus $k_1 \leq j$. If A_1 is not R, then it is M-primary and has a factorization $A_1 = A_2M^{k_2}$ with A_2 finitely generated and $k_2 \geq 1$ and necessarily with $k_1 + k_2 \leq j$. Eventually, this process must stop and yield $Q_1 = M^{k_1+k_2+\cdots+k_n} = M^j$. Thus $\{M^n\}$ is the complete set of M-primary ideals. □

In the next theorem, we show that the assumption that M is the radical of finitely generated is superfluous, but to prove this we will make use of Theorem 5.3.2. The proof is adapted from the proof of [68, Theorem 2.3].

Theorem 5.3.3. *Let R be a local domain with maximal ideal M. If R has special factorization, then M is principal.*

Proof. Assume R has special factorization, and by way of contradiction, suppose that M is not principal. For a nonzero element $t \in M$, the ideal tR factors as the product of a finitely generated ideal J and prime ideals $P_1, P_2, \ldots, P_n$. Since $tR = JP_1P_2 \cdots P_n$ is invertible and R is local, J and each P_i is principal. It follows that M is the union of the principal primes of R.

Now, let $P = pR$ be a nonzero principal prime of R, and let Q be a prime minimal over $qR + pR$ for some $q \in M \backslash P$. Then QR_Q is the radical of the finitely generated ideal $qR_Q + pR_Q$. Since R_Q has special factorization by Lemma 5.3.1, QR_Q is principal (by Theorem 5.3.2), which gives a contradiction as it properly contains the principal prime PR_Q. Hence it must be that M is principal. □

Corollary 5.3.4. *If R is a domain with special factorization, then PR_P is principal for each nonzero prime P.*

Proof. This follows from Theorem 5.3.3 in view of Lemma 5.3.1. □

The following result is immediate.

Corollary 5.3.5. *Let R be a domain with special factorization. Then for each maximal ideal M, $\{M^n\}$ is the set of M-primary ideals, $M^n \neq M^m$ for $n \neq m$ and $rR + M^n = M$ for each $r \in M \backslash M^2$.*

Combining Corollary 5.3.4 with the following general lemma leads to a simple proof that a domain with special factorization is a Prüfer domain.

Lemma 5.3.6. *Let P be a nonzero prime of a domain R. If there is a finitely generated ideal I such that P is minimal over II^{-1}, then PR_P is not principal.*

Proof. By way of contradiction, assume PR_P is principal and P is minimal over II^{-1} for some finitely generated ideal I. Let $p \in P$ be such that $pR_P = PR_P$. Then $p^n R_P = P^n R_P$ for each $n \geq 1$ and these are the only PR_P-primary ideals. Thus there is a positive integer k and a finite subset $H \subseteq I^{-1}$ such that $HIR_P = p^k R_P$. Then $p^{-k} HIR_P = R_P$. Since HI is a finitely generated ideal of R, there is an element $t \in R \backslash P$ such that $tp^{-k} HI \subseteq R$ is not contained in P. As $tp^{-k} H \subseteq I^{-1}$, we have a contradiction. □

Theorem 5.3.7. *Let R be an integral domain. If PR_P is principal for each nonzero ideal P of R, then R is a Prüfer domain. In particular, if R has special factorization, then R is a Prüfer domain.*

Proof. By Lemma 5.3.6, if some nonzero finitely generated ideal is not invertible, then there is a nonzero prime P such that PR_P is not locally principal. □

As noted by Olberding in [68, Lemma 2.1], a factorization of an ideal $I = JQ_1Q_2 \cdots Q_m$ in a Prüfer domain with $Q_1, Q_2, \ldots, Q_m$ prime (and $m \geq 1$), can be transformed into a factorization of the form $I = JP_1^{r_1} P_2^{r_2} \cdots P_n^{r_n}$ with $P_1, P_2, \ldots, P_n$ pairwise comaximal primes and each r_i a positive integer.

Corollary 5.3.8. *Let R an integral domain. Then R has special factorization if and only if each nonzero ideal I can be factored as $I = JP_1^{r_1}P_2^{r_2}\cdots P_n^{r_n}$ for some invertible ideal J and pairwise comaximal primes $P_1, P_2, \ldots, P_n$.*

Proof. If R has special factorization, it is a Prüfer domain by Theorem 5.3.7. Thus if P and Q are primes with $P \subsetneq Q$, then $QP = P$. Also, a pair of prime ideals is either comparable or comaximal. Hence any factorization of a nonzero ideal I can be "reduced" to include only powers of pairwise comaximal primes, and the finitely generated factor is invertible. □

We next show that if R has special factorization, then each nonzero nonmaximal prime is both sharp and contained in a unique maximal ideal.

Theorem 5.3.9. *If R is a (Prüfer) domain with special factorization, then each nonzero nonmaximal prime is both sharp and contained in a unique maximal ideal, which is invertible.*

Proof. Let P be nonzero nonmaximal prime, and let M be a maximal ideal that contains P. Since R is a Prüfer domain, $P^{-1} = (P : P)$ is a ring (Theorem 2.3.2(3)). By Corollary 5.3.4, we may write $PR_P = aR_P$ for some $a \in P$. Let $J := aR_M \bigcap R$. Since R has special factorization, $J = AQ_1^{r_1}Q_2^{r_2}\cdots Q_n^{r_n}$ for some invertible ideal A and pairwise comaximal prime ideals $Q_1, Q_2, \ldots, Q_n$ by Corollary 5.3.8. Since A is finitely generated, it properly contains J, and therefore $aR_M \subsetneq AR_M$. It follows that exactly one of the Q_i is contained in M, say Q_1. Then $JR_M = AQ_1^{r_1}R_M$ and, moreover, $J \subseteq AQ_1^{r_1} \subseteq AQ_1^{r_1}R_M \bigcap R = JR_M \bigcap R = J$. Thus $J = AQ_1^{r_1}$. As both $AQ_1^{r_1}R_M = JR_M = aR_M$ and MR_M are invertible, we have $Q_1 = M$ and $J = AM^r$ (with $r = r_1$).

For the remainder of the proof, we make free use of Theorem 2.5.2. There are two possibilities for J^v. If M is invertible, then $J = J^v$ is invertible, and if M is not invertible, then $(M^r)^v = R$ and $J^v = A$. Thus no matter what we know about M, J^v is invertible. A consequence of this is that J^{-1} is not a ring. On the other hand, P^{-1} is a ring that is contained in J^{-1}, so it cannot be the case that $J^v = P^v$. Hence P is sharp by the contrapositive of Theorem 2.5.2(1). Thus there is a finitely generated ideal B with $\sqrt{B} = P$ and $JR_M = BR_M$. For such an ideal B, $J^v = B(P' : P')$ where P' is the largest prime contained in all maximal ideals that contain P. Since $B \subseteq P'$ and $\mathrm{Max}(R, P) = \mathrm{Max}(R, P')$, P' is sharp. If P' is properly contained in M, then $J^{-1} = B^{-1}P'$. Since J^v is invertible, we would then have $R = J^vJ^{-1} = J^vB^{-1}P'$ and then P' would be an invertible nonmaximal prime of a Prüfer domain, which is impossible. Hence $P' = M$ and M is both sharp and the only maximal ideal that contains P. As MR_M is principal, M is invertible. □

Corollary 5.3.10. *If P is a nonzero nonmaximal prime of a (Prüfer) domain R with special factorization, then there is an invertible ideal $B \subsetneq P$ such that $\sqrt{B} = P$, $BR_P = PR_P$ and $P^n = B^{n-1}P$ for each integer $n > 1$. Moreover, P^n is divisorial for each $n \geq 1$.*

Proof. Such an invertible ideal B exists since PR_P is principal and the radical of a finitely generated ideal. Simply, let $r \in P$ be such that $rR_P = PR_P$ and let I be a finitely generated ideal with $\sqrt{I} = P$. Then $B := rR+I$ clearly satisfies $\sqrt{B} = P$ and $BR_P = PR_P$. The statement that $P^n = B^{n-1}P$ comes from checking locally. Finally, P is divisorial by [24, Corollary 4.1.12] and hence $P^n = B^{n-1}P$ is also divisorial. □

Corollary 5.3.11. *If R is a (Prüfer) domain with special factorization, then it also has weak factorization.*

Proof. By Theorems 5.3.7 and 5.3.9, R is a Prüfer domain, and each nonzero nonmaximal prime is both sharp and contained in a unique maximal ideal, and this maximal ideal is invertible. Let I be a nonzero ideal and let $I = BP_1^{r_1}P_2^{r_2}\cdots P_n^{r_n}$ be a factorization into a product of an invertible ideal B and pairwise comaximal primes $P_1, P_2, \ldots, P_n$. If P_i is either a nonmaximal prime or an invertible maximal ideal, then $P_i^{r_i}$ is divisorial (the first case following from Corollary 5.3.10). The only other possibility is that P_i is an unsteady height one maximal ideal. Split the set $\{P_1, P_2, \ldots, P_n\}$ into disjoint sets $\{Q_1, Q_2, \ldots, Q_m\}$ and $\{M_1, M_2, \ldots, M_k\}$ where the Q_i are divisorial and the M_j are unsteady (height one) maximal ideals. Rewrite the factorization as $I = BQ_1^{s_1}Q_2^{s_2}\cdots Q_m^{s_m}M_1^{t_1}M_2^{t_2}\cdots M_k^{t_k}$ (perhaps with $R = \prod_i Q_i^{s_i}$ or $R = \prod_j M_j^{t_j}$ in the event none or all of the P_i's are divisorial). As B is invertible and the P_i's are pairwise comaximal, $I^v = B(\prod_i Q_i^{s_i}\prod_j M_j^{t_j})^v = B(\prod_i (Q_i^{s_i})^v)^v = B(\bigcap_i Q_i^{s_i})^v = B(\bigcap_i Q_i^{s_i}) = B\prod Q_i^{s_i}$. Thus $I = I^v\prod_j M_j^{t_j}$. Therefore R has weak factorization. □

Lemma 5.3.12. *If R is a (Prüfer) domain with special factorization, then each overring has both special factorization and weak factorization.*

Proof. Assume that R has special factorization, and let T be an overring of R. By Corollary 5.3.11, it suffices to prove that T has special factorization. Let J be a nonzero ideal of T. Then $J = IT$, where $I := J \bigcap R$. Factor I as $I = AP_1P_2\cdots P_n$ with A invertible and $P_1, P_2, \ldots, P_n$ primes of R, not necessarily distinct but with $n \geq 1$. Since A is invertible, it properly contains I. Hence AT properly contains $IT = J$. It follows that at least one of the primes P_i survives in T and therefore extends to a prime of T. Renumber if necessary to have $P_iT \neq T$ for $1 \leq i \leq m$ and $P_jT = T$ for $m < j$ (possibly with $m = n$). The factorization $J = AP_1P_2\cdots P_mT$ has AT invertible and prime ideals $P_1T, P_2T, \ldots, P_mT$, with $m \geq 1$. □

Theorem 5.3.13. *The following statements are equivalent for an integral domain R.*

(i) R has special factorization.
(ii) R is an h-local generalized Dedekind domain.
(iii) R has strong pseudo-Dedekind factorization.
(iv) R has pseudo-Dedekind factorization, and each maximal ideal of R is invertible.

Proof. It is clear that (iii) implies (i). It is almost as clear that (iv) implies (iii). The factorization of a nonzero noninvertible ideal is taken care of by pseudo-Dedekind factorization. For an invertible ideal B, we have $B \subsetneq M$ for some maximal ideal M. Then BM^{-1} is invertible (equal to R if $B = M$) and $B = (BM^{-1})M$ since M is invertible. Hence (iv) implies (iii).

To see that (i) implies both (ii) and (iv), first recall from Theorem 5.3.13 and Corollary 5.3.4 that a domain with special factorization has weak factorization and no nonzero idempotent primes. The main step in this proof is to show that R has no unsteady maximal ideals. By way of contradiction, assume M is an unsteady maximal ideal. By finite unsteady character (Theorem 4.2.10), there is a finitely generated ideal I that is contained in M and no other unsteady maximal ideal. Since $\mathrm{Spec}(R)$ is treed and each nonzero prime is contained in a unique maximal ideal (Theorem 5.3.9), I must have infinitely many minimal primes other than M. By Lemma 4.3.6, the maximal ideals of $\Phi(I)$ are all extended from minimal primes of I in R. Also, since $\Phi(I)$ has both special factorization and weak factorization (Lemma 5.3.12), we may assume $R = \Phi(I)$ with $\sqrt{I}$ the Jacobson radical of R. Let $\{M_\alpha\}$ be the other minimal primes of I. Then each M_α is an invertible maximal ideal of R, and there are infinitely many such ideals (otherwise, the original M is the radical of a finitely generated ideal).

Let J be the Jacobson radical of R and factor it as $J = CM_1M_2\cdots M_n$ with C invertible and $M_1, M_2, \ldots, M_n$ maximal ideals of R. Since J is a radical ideal and no maximal ideal is idempotent, the M_i are distinct. Note that since M is the only maximal ideal that is not sharp, J is the intersection of the invertible maximal ideals (by Theorem 2.5.10). If J is not invertible, then we may assume $M_1 = M$ and $n = 1$, but this puts C in each invertible maximal ideal and we have $C = J = CM$, which is impossible as C is invertible. Thus J is invertible and M is not one of the M_i.

As $\mathrm{Max}(R)$ is an infinite set, we may partition $\mathrm{Max}(R)\backslash\{M\}$ into two infinite sets $\{Q_\gamma\}$ and $\{N_\beta\}$. Let $Q := \bigcap_\gamma Q_\gamma$ and $N := \bigcap_\beta N_\beta$. By Theorem 2.5.6, no N_β contains Q and no Q_γ contains N. Thus both properly contain J. On the other hand, each Q_γ is invertible and minimal over Q and each N_β is invertible and minimal over N. Also, by Lemma 2.5.1, no N_β survives in the ring $\Gamma(Q)$ and no Q_γ survives in $\Gamma(N)$. Since $\Gamma(Q)$ has infinitely many invertible maximal ideals, each minimal over the Jacobson radical, it must have at least one noninvertible maximal ideal (Theorem 2.5.11). The only possibility is $M\Gamma(Q)$. Moreover, $M\Gamma(Q)$ is the unique noninvertible maximal ideal of $\Gamma(Q)$. Similarly, $M\Gamma(N)$ is the unique noninvertible maximal ideal of $\Gamma(N)$. Therefore M must contain both Q and N (Lemma 2.5.1). But, as with J, both Q and N are invertible, otherwise, $Q = AM$ and/or $N = BM$ for invertible ideals $A \subseteq \bigcap_\gamma Q_\gamma$ and/or $B \subseteq \bigcap_\beta N_\beta$. Obviously, M is the only prime ideal that contains $Q + N$, but $Q + N$ is the sum of two finitely generated ideals which would make M sharp, a contradiction. Hence R has no unsteady maximal ideals. Thus by Theorem 4.4.8, R is an h-local Prüfer domain such that each maximal ideal is invertible.

To see that R has pseudo-Dedekind factorization, revisit the proof that a domain with special factorization has weak factorization (Corollary 5.3.11). For the ideal

$I = BP_1^{r_1} P^{r_2} \cdots P_n^{r_n}$ with B invertible and the P_i pairwise comaximal, each P_i is a sharp prime that is not idempotent. Hence for each P_i there is an invertible ideal E_i with $\sqrt{E_i} = P_i$ and $P_i R_{P_i} = E_i R_{P_i}$. It follows that $P_i^{r_i} = E_i^{r_i - 1} P_i$. Substituting into the original factorization, we have $I = BEP_1 P_2 \cdots P_n$ with both B and $E := \prod_i E_i^{r_i-1}$ invertible. Thus R has pseudo-Dedekind factorization.

There are several ways to conclude that R is also a generalized Dedekind domain. For example, PR_P is principal for each nonzero prime P, but P is sharp (so the radical of a finitely generated ideal). So, R is a generalized Dedekind domain by Theorem 3.3.2. For an alternate proof, use pseudo-Dedekind factorization to see that each divisorial ideal factors as the product of an invertible ideal and a finite product of pairwise comaximal primes. Each nonzero prime is divisorial, and thus so is an arbitrary finite product of comaximal primes. Apply Theorem 3.3.6 to see that R is a generalized Dedekind domain.

To finish the proof, we show that (ii) implies (iv). Suppose R is an h-local generalized Dedekind domain. From the generalized Dedekind assumption, each PR_P is principal for each nonzero prime P and each such P is sharp (Theorem 3.3.2). In particular, each maximal ideal is invertible. For each maximal ideal M, R_M has pseudo-Dedekind factorization (Theorem 5.2.9) which combines with the h-local assumption to yield that R has pseudo-Dedekind factorization with each maximal ideal invertible (Theorem 5.1.8). □

5.4 Factorization and the Ring $R(X)$

Let R be an integral domain and let X be an indeterminate over R. Recall that, for each $h \in R[X]$, $\mathbf{c}(h)$ denotes the content of the polynomial h; i.e., the ideal of R generated by the coefficient of h. The *Nagata ring* of R is the ring $R(X) := R[X]_{\mathscr{U}}$ where $\mathscr{U} := \{h \in R[X] \mid \mathbf{c}(h) = R\}$.

We may extend the notion of the content of a single polynomial to ideals of both $R[X]$ and $R(X)$. For a nonzero ideal I of $R[X]$, the *content* of I is the ideal of R generated by the coefficients of the polynomials contained in I. It is quite easy to show that the content of I is the union $\mathbf{c}(I) := \bigcup\{\mathbf{c}(g) \mid g \in I\}$. A similar definition can be used with regard to ideals of $R(X)$. For each nonzero ideal J of $R(X)$, the ideal $I_J := J \bigcap R[X]$ is such that $J = I_J R(X)$. Using this notation, we set $\mathbf{c}(J) := \mathbf{c}(I_J)$, and again refer to $\mathbf{c}(J)$ as the *content* of J.

It is well-known that if R is a Prüfer domain, then not only is $R(X)$ a Bézout domain but it coincides with the Kronecker function ring of R [34, Theorems 32.7 and 33.4]. Hence $gR(X) = \mathbf{c}(g)R(X)$ for each nonzero $g \in R[X]$ (for an explicit proof that $gR(X) = \mathbf{c}(g)R(X)$ see the proof of [34, Theorem 32.7]). Moreover, if V is a valuation domain, then $V(X)$ is the trivial extension of V in the field of rational functions $K(X)$ [34, Propositions 18.7 and 33.1].

The following lemma collects several other useful facts about $R(X)$ when R is a Prüfer domain.

Lemma 5.4.1. *Let R a Prüfer domain.*

(1) *For each nonzero ideal I of R, $I^v R(X) = (IR(X))^v$.*
(2) *Each ideal of $R(X)$ is extended from an ideal of R. Specifically, if J is an ideal of $R(X)$, then $J = \mathbf{c}(J)R(X)$.*
(3) *The map $I \mapsto IR(X)$ establishes an order-preserving bijection between the set of the invertible (resp., divisorial, prime, primary, maximal) ideals of R and the set of the invertible (resp., divisorial, prime, primary, maximal) ideals of $R(X)$.*
(4) *R has finite character if and only if $R(X)$ has finite character.*
(5) *Each nonzero prime ideal in contained in a unique maximal ideal in R if and only if the same is true in $R(X)$.*
(6) *R is h-local if and only if $R(X)$ is h-local.*

Proof. The statement in (1) follows from the equality $(R : I)R(X) = (R(X) : IR(X))$ that holds for each nonzero ideal I of an arbitrary integral domain R by [20, Lemma 4.2] and its proof. The statement in (2) follows from the fact that $gR(X) = \mathbf{c}(g)R(X)$ for each polynomial $g \in R[X]$. For (3), use (1), (2) and [34, Proposition 33.1]. Both (4) and (5) follow directly from (3), and (6) follows from (4) and (5). □

Using this lemma, it is relatively easy to see that R and $R(X)$ have similar factoring properties when R is a Prüfer domain. For example, if I is a nonzero nondivisorial ideal of R that can be factored as $I = I^v P_1 P_2 \cdots P_n$ where each P_i is a prime (maximal) ideal of R, then $IR(X) = (IR(X))^v \prod(P_i R(X))$ with each $P_i R(X)$ a prime (maximal) ideal of $R(X)$. If, in addition, the P_i's are exactly the nondivisorial maximal ideals M of R for which IR_M is not divisorial in R_M, then the $P_i R(X)$'s play the same role in $R(X)$. Also, if no P_i can be omitted, then the same is true about the $P_i R(X)$'s. As a first application of Lemma 5.4.1, we can avoid dealing specifically with factoring individual ideals by making use of statement (6) and Theorem 4.1.2.

Corollary 5.4.2. *A Prüfer domain R has the (very) strong factorization property if and only if $R(X)$ has the same property.*

At this point, it is natural to show explicitly the factorization of a nonzero nondivisorial ideal J of $R(X)$, when R is Prüfer with the weak factorization property. After that we show how to factor such an ideal J when R is Prüfer with the (very) strong factorization property

Theorem 5.4.3. *Let R be a Prüfer domain.*

(1) *If R has the weak factorization property and if J is a nonzero nondivisorial ideal of $R(X)$, then $J = J^v \prod_i N_i^{s_i}$, with $J^v = \mathbf{c}(J)^v R(X)$ and $N_i = M_i R(X)$, where the M_i's are exactly the maximal ideals of R appearing in the weak factorization $\mathbf{c}(J) = \mathbf{c}(J)^v \prod_i M_i^{s_i}$ in the domain R.*
(2) *If $R(X)$ has the weak factorization property and if I is a nonzero nondivisorial ideal of R, then $IR(X) = I^v(X) \prod_i (M_i R(X))^{r_i}$ for an appropriate finite family of maximal ideals $M_i R(X)$'s of $R(X)$ and so, by intersect-*

ing with R, we obtain $I = I^v \prod_i M_i^{r_i}$ *(note that* $I^v R(X) \prod_i (M_i R(X))^{r_i} = (I^v \prod_i M_i^{r_i}) R(X)$*).*

Proof. The factorization of the ideal $IR(X)$ in statement (2) follows from Lemma 5.4.1 and the discussion above. To establish (1), assume R has the weak factorization property and let J be a nonzero nondivisorial ideal of $R(X)$. Then by Lemma 5.4.1, we have both $J = \mathbf{c}(J)R(X)$ and $J^v = \mathbf{c}(J)^v R(X)$. Since R has the weak factorization property, $\mathbf{c}(J) = \mathbf{c}(J)^v M_1^{s_1} M_2^{s_2} \cdots M_n^{s_n}$ for some maximal ideals $M_1, M_2, \cdots, M_n$ of R. Another application of Lemma 5.4.1, yields a weak factorization of J as $J = J^v \prod N_i^{s_i}$ with $N_i = M_i R(X)$ for each i. □

Corollary 5.4.4. *Let R be a Prüfer domain. Then R has the weak factorization property if and only if R(X) has the weak factorization property.*

For any domain R, the maximal ideals of $R(X)$ are the ideals of the form $MR(X)$ where M ranges over the set of maximal ideals of R. Also $R_M(X) = R(X)_{MR(X)}$ for each M. In the case R is a Prüfer domain, if J is an ideal of $R(X)$ and $\mathbf{c}(J)$ ($\neq \mathbf{c}(J)^v$) factors as $\mathbf{c}(J) = \mathbf{c}(J)^v M_1 M_2 \cdots M_n$ where the M_i's are the distinct maximal ideals M of R such that $\mathbf{c}(J)R_M$ is not divisorial, then $J = J^v N_1 N_2 \cdots N_n$ with each $N_i = M_i R(X)$ a maximal ideal of $R(X)$ such that $JR(X)_{N_i}$ is not divisorial (and $JR(X)_N$ divisorial for all other maximal ideals N of $R(X)$).

Since it is easy to verify that an overring of an h-local Prüfer domain is again h-local [27, Chap. IV, Proposition 3.16], then an overring of a Prüfer domain with strong factorization has strong factorization by Theorem 4.4.8. On the other hand, recall that not all overrings of a Prüfer domain with weak factorization have weak factorization (Example 4.3.4). The previous observations show that, for the problem of studying the overrings of a Prüfer domain with the weak factorization property (see Corollary 4.3.2), we can assume (without loss of generality) that the domain is a Bézout domain.

For pseudo-Dedekind factorization, it is relatively easy to show that if R is an integrally closed domain with pseudo-Dedekind factorization, then $R(X)$ has pseudo-Dedekind factorization. The converse is also valid. Moreover, while more complicated to prove, it is also the case that for any domain R, R has pseudo-Dedekind factorization if and only if $R(X)$ has pseudo-Dedekind factorization.

Lemma 5.4.5. *For a domain R, R is an h-local domain such that its integral closure R′ is a Prüfer domain if and only if R(X) is h-local with integral closure R′(X) a Prüfer domain.*

Proof. For any domain R, each maximal ideal of $R(X)$ is of the form $MR(X)$ for some maximal ideal M of R. It follows that if $f = g/u$ is a nonzero nonunit of $R(X)$ with $g, u \in R[X]$ and $\mathbf{c}(u) = R$, then the only maximal ideals of $R(X)$ that contain f are those that contain $\mathbf{c}(g)$. Hence f is contained in only finitely many maximal ideals of $R(X)$ if and only if $\mathbf{c}(g)$ is contained in only finitely many maximal ideals of R. Thus R has finite character if and only if $R(X)$ has finite character.

By [37, Theorem 3], $R'(X)$ is the integral closure of $R(X)$ (no matter whether R' is Prüfer or not). Combining [34, Theorem 33.4] and [5, Theorem 2.7], we have that the following are equivalent: (i) each prime ideal of $R(X)$ is extended from a prime ideal of R, (ii) R' is a Prüfer domain, (iii) $R'(X)$ is a Prüfer domain. Therefore R is h-local with R' a Prüfer domain if and only if $R(X)$ is h-local with $R'(X)$ a Prüfer domain. □

For any domain R, if I is a nonzero ideal of R, then $IR(X) \bigcap R = I$ [50, Theorem 14.1(3)]. Thus if there are ideals $J_1, J_2, \ldots, J_n$ of R such that $IR(X) = J_1 J_2 \cdots J_n R(X)$, then $I = J_1 J_2 \cdots J_n$.

Theorem 5.4.6. *Let R be an integrally closed domain. Then R has pseudo-Dedekind factorization if and only if $R(X)$ has pseudo-Dedekind factorization.*

Proof. An integrally closed domain that has pseudo-Dedekind factorization is an h-local Prüfer domain (Corollary 5.1.7). Also by [34, Theorem 34.2], R is a Prüfer domain if and only if $R(X)$ is a Prüfer domain. Hence if either R or $R(X)$ has pseudo-Dedekind factorization, then each is a Prüfer domain. In this case, each ideal of $R(X)$ is extended from an ideal of R. In particular, each prime ideal of $R(X)$ is extended from a prime ideal of R and each invertible ideal of $R(X)$ is extended from an invertible ideal of R. Thus factorizations in R extend to factorizations in $R(X)$ and factorizations in $R(X)$ contract to factorizations in R. Hence $R(X)$ has pseudo-Dedekind factorization if and only if R has pseudo-Dedekind factorization. □

Essentially the same proof is valid for a domain with strong pseudo-Dedekind factorization (equivalently, special factorization). In this case, we do not need to assume, a priori, that R is integrally closed as it must be a Prüfer domain by Theorem 5.3.13.

Theorem 5.4.7. *For a domain R, R is a has strong pseudo-Dedekind factorization if and only if $R(X)$ has strong pseudo-Dedekind factorization.*

Theorem 5.4.8. *For a domain R, R has pseudo-Dedekind factorization if and only if $R(X)$ has pseudo-Dedekind factorization.*

Proof. By Theorem 5.4.6, we may assume R is not integrally closed. It is well-known that $\mathrm{Max}(R(X)) = \{MR(X) \mid M \in \mathrm{Max}(R)\}$. Moreover, for each maximal ideal M of R, $R_M(X) = R(X)_{MR(X)}$.

Assume R has pseudo-Dedekind factorization. Then by Corollary 5.2.14, R is h-local and for each maximal ideal M, R_M is a PVD with pseudo-Dedekind factorization. We also have that R' is a Prüfer domain with pseudo-Dedekind factorization (Theorem 5.2.13) and that $R(X)$ is h-local with integral closure $R'(X)$ (Lemma 5.4.5). Moreover, MR' is the unique maximal ideal of R' that lies over M, and the integral closure of R_M is the valuation domain R'_M. If $R_M \doteq R'_M$, then $R_M(X) = R(X)_{MR(X)}$ is a valuation domain with pseudo-Dedekind factorization by Theorem 5.4.6. By Theorem 5.2.4, the other possibility is that $[R'/MR' : R/M] = 2$. In this case, we also have $[R'_M(X)/MR'_M(X) : R_M(X)/MR_M(X)] = 2$. Hence we always have that $R_M(X)$ is a PVD with pseudo-Dedekind factorization. Thus $R(X)$ has pseudo-Dedekind factorization (Corollary 5.2.14).

For the converse, suppose $R(X)$ has pseudo-Dedekind factorization. Then $R(X)$ is h-local and $R_M(X)$ is a PVD with pseudo-Dedekind factorization for each maximal ideal M of R (Corollary 5.2.14). Also $R'(X)$ is a Prüfer domain with pseudo-Dedekind factorization (Theorem 5.2.13). Therefore by [5, Theorem 2.7], each prime ideal of $R(X)$ is extended from a prime of R. By Lemma 5.4.5, R is h-local and R' is a Prüfer domain. Thus by Corollary 5.2.14, it suffices to show that R_M is a PVD with pseudo-Dedekind factorization for each $M \in \mathrm{Max}(R)$. If R_M is a valuation domain, then we simply invoke Theorem 5.4.6. If R_M is not a valuation domain, then neither is $R_M(X)$. Hence in this case, $R'_M(X)$ is the canonical valuation overring of $R_M(X)$ and $[R'_M(X)/MR'_M(X) : R_M(X)/MR_M(X)] = 2$. It follows that R'_M is the canonical valuation overring of R_M with $[R'_M/MR'_M : R_M/MR_M] = 2$. Therefore R has pseudo-Dedekind factorization. □

One might be tempted to try the following approach to establishing the implication that R has pseudo-Dedekind factorization whenever $R(X)$ does. For a nonzero noninvertible ideal I of R, factor the extension $IR(X) = AP_1P_2\cdots P_n$ for some invertible ideal A of $R(X)$ and prime ideals $P_1, P_2, \ldots, P_n$ (each prime of $R(X)$ is extended since $R'(X)$ is a Prüfer domain). Then simply "show" that $A = JR(X)$ for some invertible ideal J of R. However, while there may be an invertible ideal J such that $IR(X) = JP_1P_2\cdots P_nR(X)$, there may be no such J that extends to the ideal A. For example, let $R := \mathbb{Q} + Y\mathbb{Q}[\sqrt{2}][[Y]]$ (with $M := Y\mathbb{Q}[\sqrt{2}][[Y]]$) and let $f(X) := YX + \sqrt{2}Y$. Then R has pseudo-Dedekind factorization (it is a PVD with integral closure $R' = \mathbb{Q}[\sqrt{2}][[Y]]$ a valuation domain such that $[R'/M : R/M] = 2$). Since $\sqrt{2}$ is not in R, $f(X)$ does not factor as $Y(X + \sqrt{2})$ in $R[X]$. However, $f(X)(YX - \sqrt{2}Y) = Y^2(X^2 - 2)$ and $f(X)(\sqrt{2}YX - 2Y) = \sqrt{2}Y(X^2 - 2)$ are valid factorizations in $R[X]$. Thus $Y^2, \sqrt{2}Y^2 \in f(X)R(X)$. In fact, $f(X)R(X) \bigcap R = M^2$. We know $M^2 = YM$ is a valid factorization in R, and thus $M^2R(X) = YMR(X)$. However, we also have that $M^2R(X) = f(X)MR(X)$ is a pseudo-Dedekind factorization of $M^2R(X)$ in $R(X)$, but in this factorization $f(X)R(X)$ is not extended from an (invertible) ideal of R.

Chapter 6
Factorization and Intersections of Overrings

Abstract In the first section, we introduce the notion of an h-local maximal ideal as a maximal ideal M of a domain R such that $\Theta(M)R_M = K$ (the quotient field of R). The second section deals with independent pairs of overrings of a domain R. In the case R can be realized as the intersection of a pair of independent overrings, we show that R shares various factorization properties with these overrings. For example, R has weak factorization if and only if both overrings have weak factorization. The third section introduces Jaffard families and Matlis partitions. Just as domains of Dedekind type are the same as h-local domains, a domain R can be realized as an intersection of the domains of a Jaffard family if and only if its set of maximal ideals can be partitioned into a Matlis partition (definitions below). As in the second section, if $R = \bigcap_{\alpha \in \mathscr{A}} S_\alpha$ where $\{S_\alpha\}_{\alpha \in \mathscr{A}}$ is a Jaffard family, then R satisfies a particular factoring property if and only if each S_α satisfies the same factoring property. The last section is devoted to constructing examples using various Jaffard families.

6.1 h-Local Maximal Ideals

We assume in the present section that R is an integral domain with quotient field $K \neq R$.

Lemma 6.1.1. *Let M be a maximal ideal of an integral domain R and let $I \subseteq M$ be a nonzero ideal. Then each minimal prime of $A := IR_M \bigcap R$ is contained in M.*

Proof. First, we observe that $\sqrt{A} = \sqrt{I}\, R_M \bigcap R$. In fact, $b^n \in A$ implies $b^n s \in I$ for some $s \in R \backslash M$. It follows that $b^n s^n \in I$, so $bs \in \sqrt{I}$. Thus $\sqrt{A} \subseteq \sqrt{I}\, R_M \bigcap R$. For the reverse containment, $c \in \sqrt{I}\, R_M \bigcap R$ implies $ct \in \sqrt{I}$ some $t \in R \backslash M$. Hence $c^m t^m \in I$ some m. Therefore $c^m \in A$.

Thus we may assume I is a radical ideal. By way of contradiction suppose some minimal prime Q of A is comaximal with M. Then there are elements $r \in Q$ and

M. Fontana et al., *Factoring Ideals in Integral Domains*, Lecture Notes of the Unione Matematica Italiana 14, DOI 10.1007/978-3-642-31712-5_6,

$m \in M$ such that $r + m = 1$. Since Q is minimal over $A, AR_Q = QR_Q$. Hence there is an element $t \in R \setminus Q$ such that $rt \in A$. This puts $t \in M$ and $t = rt/r \in AR_M$ and thus we have $t \in A$. But this implies $t \in Q$ a contradiction. □

Recall that a Prüfer domain R with weak factorization also has aRTP (Theorem 4.2.4): for each nonzero noninvertible ideal I, $II^{-1}R_M$ is a radical ideal whenever M is either a steady maximal ideal or an unsteady maximal ideal that is not minimal over II^{-1} (as unsteady maximal ideals have height one when R is Prüfer with weak factorization, the second case occurs only when M does not contain II^{-1}).

Lemma 6.1.2. *Let R be a Prüfer domain with weak factorization and let M be an invertible maximal ideal of R. Then for each nonzero ideal $I \subseteq M$, $A := IR_M \bigcap R$ is divisorial.*

Proof. Let I be a nonzero ideal that is contained in M and let $A := IR_M \cap R$. Since R has weak factorization, each nonzero prime ideal P that is contained in M is contained in no other maximal ideal of R (Lemma 4.2.3). Thus by Lemma 6.1.1, M is the only maximal ideal that contains A. Hence the only possible factorization for A is as $A = A^v M^n$ for some nonnegative integer n. As M is invertible, $A^v M^n$ is divisorial, and thus we have $A = A^v$. □

Recall that a domain R is h-local if each nonzero prime ideal is contained in a unique maximal ideal and each nonzero nonunit is contained in only finitely many maximal ideals. This is equivalent to having $\Theta(M)R_M = K$ for each maximal ideal M [60, Theorem 8.5]. With this equivalence in mind, we say that a maximal ideal M of R is an *h-local maximal ideal of* R if $\Theta(M)R_M = K$. Note that if P is a prime ideal of R that is contained in M and at least one other maximal ideal N, then $R_P \supseteq R_M$ and $R_P \supseteq R_N \supseteq \Theta(M)$. Hence $P = (0)$. Thus an h-local maximal ideal M has the property that each nonzero prime ideal $Q \subseteq M$ is such that $\text{Max}(R, Q) = \{M\}$.

A family $\mathscr{F} = \{P_\alpha\}_{\alpha \in \mathscr{A}}$ of nonzero prime ideals is said to be *defining family* of R if $R = \bigcap R_{P_\alpha}$. In [4], Anderson and Zafrullah generalized the notion of an h-local domain and in the process introduced the notion of a given prime P in a defining family $\mathscr{F}$ being *$\mathscr{F}$-independent* if no nonzero prime ideal that is contained in P is contained in some other prime $Q \in \mathscr{F}$; equivalently for each $Q \in \mathscr{F} \setminus \{P\}$, no nonzero prime ideal is contained in $P \bigcap Q$. For example, if we take $\mathscr{F} := \text{Max}(R)$, then each h-local maximal ideal is $\mathscr{F}$-independent. The converse does not hold as a non-invertible maximal ideal of an almost Dedekind domain is not an h-local maximal ideal, but trivially each maximal ideal is $\mathscr{F}$-independent. In Theorems 6.1.3 and 6.1.4 we show that an h-local maximal ideal has many of the same properties of each maximal ideal of an h-local domain and is itself characterized as h-local in ways similar to several of the various properties which provide global characterizations of h-local domains.

Theorem 6.1.3. *Let R be a domain and let M be an h-local maximal ideal of R.*

(1) If P is a nonzero prime ideal that is contained in M, then $P\Theta(M) = \Theta(M)$.
(2) If I is a nonzero ideal such that $\text{Max}(R, I) = \{M\}$, *then*

(a) $I\Theta(M) = \Theta(M)$,
(b) $(R : I)R_M = (R_M : IR_M)$,
(c) $(IR_M)^v = I^v R_M$, *and*
(d) I *is divisorial if and only if* IR_M *is divisorial.*

(3) $\Theta(M)$ *is R-flat.*
(4) If I is an ideal that is comaximal with M, then

(a) $(R : I)\Theta(M) = (\Theta(M) : I\Theta(M))$,
(b) $I^v = (I\Theta(M))^v$, *and*
(c) I *is divisorial if and only if* $I\Theta(M)$ *is divisorial.*

(5) If J is a nonzero ideal of R, then $JR_M \bigcap R$ and $J\Theta(M) \bigcap R$ are comaximal.

Proof. If R is local, then $\Theta(M)$ is the quotient field of R. In this case, all of the conclusions are trivial. So, we may assume R has at least two maximal ideals.

Let P be a nonzero prime ideal that is contained in M. By way of contradiction, assume $P\Theta(M) \neq \Theta(M)$. Then there is a prime ideal Q' of $\Theta(M)$ such that $Q' \supseteq P\Theta(M)$. Let $Q := Q' \bigcap R$. Then $Q \supseteq P$ and we have $R_P \supseteq R_Q$. Also $R_Q \subseteq \Theta(M)_{Q'}$ since $R \backslash Q \subseteq \Theta(M) \backslash Q'$. From the discussion above, we have $\text{Max}(R, P) = \text{Max}(R, Q) = \{M\}$. But then we have $R_M \subseteq R_Q \subseteq \Theta(M)_{Q'} \subsetneq K = \Theta(M)R_M$, a contradiction. Therefore $P\Theta(M) = \Theta(M)$.

Let I be a nonzero ideal of R. First we consider the case that M is the only maximal ideal that contains I. Let $H := (IR_M)^v \bigcap R$. Then $HR_M = (IR_M)^v$. Since M is the only maximal ideal that contains I, each prime ideal that contains I blows up in $\Theta(M)$. Hence we also have $I\Theta(M) = \Theta(M)$. For duals we have $(R : H) = (R_M : HR_M) \bigcap \Theta(M) = (R_M : IR_M) \bigcap \Theta(M) = (R : I)$. Hence $I \subseteq H \subseteq I^v$.

Since $R = R_M \bigcap \Theta(M)$, we have $(R : I) = (R_M : IR_M) \bigcap (\Theta(M) : I\Theta(M)) = (R_M : IR_M) \bigcap \Theta(M)$. Multiplying by R_M yields $(R : I)R_M = (R_M : IR_M) \bigcap \Theta(M)R_M = (R_M : IR_M)$ (since $\Theta(M)R_M$ is the quotient field of R). Hence $(IR_M)^v = (I^v R_M)^v$ and therefore $H = I^v$ with $I^v R_M$ a divisorial ideal of R_M.

To see that $\Theta(M)$ is flat it suffices to show $(R :_R t)\Theta(M) = \Theta(M)$ for each $t \in \Theta(M)$. This is quite simple. It is trivial for $t \in R$ and for $t \in \Theta(M) \backslash R$, the fact that $t \in R_N$ for each maximal ideal $N \neq M$, implies M is the only maximal ideal of R that contains $(R :_R t)$. Hence $(R :_R t)\Theta(M) = \Theta(M)$ by (1).

For (4), we now assume $I + M = R$ and let $H := (I\Theta(M))^v \bigcap R$. Since $\Theta(M)$ is R-flat, $H\Theta(M) = (I\Theta(M))^v$. Also $IR_M = HR_M = R_M$ and therefore $(\Theta(M) : I) = (\Theta(M) : H)$ and $(R : I) = R_M \bigcap (\Theta(M) : I) = R_M \bigcap (\Theta(M) : H) = (R : H)$. So, as in (1), $I \subseteq H \subseteq I^v$. Again, taking advantage of having $\Theta(M)R_M = K$, we obtain $(R : I)\Theta(M) = (R_M \bigcap (\Theta(M) : I))\Theta(M) = (\Theta(M) : I)$. Similarly, $(R : I)\Theta(M) = (R : I^v)\Theta(M) = (\Theta(M) : I^v)$. Thus $H = I^v$ and $(I^v\Theta(M))^v = (I\Theta(M))^v = I^v\Theta(M)$.

For (5), let $A := JR_M \bigcap R$ and $B := J\Theta(M) \bigcap R$. There is nothing to prove if either J is comaximal with M or $\text{Max}(R, J) = \{M\}$. So we may assume both A and B are proper ideals. From above, we know $\text{Max}(R, A) = \{M\}$. So, it suffices to show $B + M = R$. By way of contradiction, assume $B \subseteq M$ and let $P \subseteq M$ be a minimal prime of B. Since $P\Theta(M) = \Theta(M)$, there is a finitely generated ideal $G \subseteq P$ such that $G\Theta(M) = \Theta(M)$. It follows that BR_P contains a power of GR_P. Since G is finitely generated, there is an element $t \in R \backslash P$ such that $tG^n \subseteq B$. As $G^n\Theta(M) = \Theta(M)$, we have $t \in B\Theta(M) \bigcap R = B$, contradicting the assumption that P contains B. Hence $B + M = R$. □

Theorem 6.1.4. *Let R be a domain that is not local and let M be a maximal ideal of R. Then the following are equivalent.*

(i) M is an h-local maximal ideal of R.
(ii) $P\Theta(M) = \Theta(M)$ for each nonzero prime ideal $P \subseteq M$.
(iii) $R_M R_N = K$ for each maximal ideal $N \neq M$.
(iv) Each nonzero ideal of $\Theta(M)$ is extended from an ideal of R that is comaximal with M.
(v) Each nonzero prime ideal of $\Theta(M)$ is extended from a prime ideal of R that is comaximal with M.
(vi) For each nonzero ideal I of R, $A := IR_M \bigcap R$ and $B := I\Theta(M) \bigcap R$ are comaximal.
(vii) For each nonzero prime ideal $P \subseteq M$, $\text{Max}(R, P) = \{M\}$ and there is an invertible ideal $J \subseteq P$ such that $\text{Max}(R, J) = \{M\}$.

Proof. It is clear that (iv) implies (v). Note that if P is a nonzero prime ideal that is contained in a maximal ideal N different from M, then $R_P \supseteq R_N \supseteq \Theta(M)$. Hence $P\Theta(M)$ is a proper ideal of $\Theta(M)$.

To see that (i) implies (ii), let P be a nonzero prime ideal that is contained in M. By way of contradiction, assume $P\Theta(M) \neq \Theta(M)$. Then there is a prime ideal Q' of $\Theta(M)$ such that $Q' \supseteq P\Theta(M)$. Let $Q := Q' \bigcap R$. Then $Q \supseteq P$ and we have $R_P \supseteq R_Q$. Also $R_Q \subseteq \Theta(M)_{Q'}$ since $R \backslash Q \subseteq \Theta(M) \backslash Q'$. From the discussion above, we have $\text{Max}(R, P) = \text{Max}(R, Q) = \{M\}$. But then we have $R_M \subseteq R_Q \subseteq \Theta(M)_{Q'} \subsetneq K = \Theta(M)R_M$, a contradiction. Therefore $P\Theta(M) = \Theta(M)$. Note we also have that $\text{Max}(R, P) = \{M\}$.

(ii) $\Rightarrow$ (i) We prove the contrapositive. Suppose $\Theta(M)R_M$ is a proper subring of K. Then there is a valuation domain $V(\neq K)$ that contains $\Theta(M)R_M$. Let $P := N \bigcap R$ where N is the maximal ideal of V. Since $V \supseteq R_M$, $P \subseteq M$ and certainly $P\Theta(M)$ is a proper ideal of $\Theta(M)$.

For (v) $\Rightarrow$ (i), we prove the contrapositive. As in the proof of (ii) implies (i), assume $\Theta(M)R_M$ is a proper subring of K and let $V(\neq K)$ be a valuation domain that contains $\Theta(M)R_M$. Also let $P' := N \bigcap \Theta(M)$ and $P := N \bigcap R$, where N is the maximal ideal of V. We have $P' \supseteq P\Theta(M)$ and $P \subseteq M$. It follows that there is no prime ideal Q of R that is comaximal with M and extends to P' in $\Theta(M)$.

(i) $\Rightarrow$ (iv) & (vi) By Theorem 6.1.3, $\Theta(M)$ is R-flat. Hence each ideal of $\Theta(M)$ is extended from an ideal of R. Let I be a nonzero ideal of R and let $A := IR_M \bigcap R$

and $B := I\Theta(M) \bigcap R$. By Lemma 6.1.1, each minimal prime of A is contained in M. We first show that $A\Theta(M) = \Theta(M)$. By way of contradiction, suppose there is a prime ideal Q' of $\Theta(M)$ such that $Q' \supseteq A\Theta(M)$. Then $Q := Q' \bigcap R$ is a prime ideal that contains A. Next, let $P \subseteq Q$ be a minimal prime of A. Then $P \subseteq M$ which then implies $Q' \supseteq P\Theta(M) = \Theta(M)$, a contradiction. Hence $A\Theta(M) = \Theta(M)$. We also have that each prime ideal that contains A is contained in M and no other maximal ideal.

To see that A and B are comaximal it suffices to show $B + M = R$. By way of contradiction, assume $B \subseteq M$ and let $P \subseteq M$ be a minimal prime of B. Since $P\Theta(M) = \Theta(M)$, there is a finitely generated ideal $G \subseteq P$ such that $G\Theta(M) = \Theta(M)$. Thus there is an element $t \in R \backslash P$ such that $tG^n \subseteq B$ for some positive integer n. It follows that $tG^n\Theta(M) = t\Theta(M) \subseteq B\Theta(M)$ and from this we deduce that $t \in B$, a contradiction. Thus $B + M = R$ and we also have $B + A = R$.

(vi) $\Rightarrow$ (v) Assume that for each nonzero ideal I of R, $IR_M \bigcap R$ and $I\Theta(M) \bigcap R$ are comaximal. We first show that $\Theta(M)$ is R-flat. For this it suffices to show that $(R :_R t)\Theta(M) = \Theta(M)$ for each $t \in \Theta(M)\backslash R$. Fix $t \in \Theta(M)\backslash R$. Since $t \in R_N$ for each maximal ideal $N \neq M$, M is the only maximal ideal that contains $J := (R :_R t)$. The ideal $J\Theta(M) \bigcap R$ contains J and is comaximal with $JR_M \bigcap R$. Hence $J\Theta(M) = \Theta(M)$ and therefore $\Theta(M)$ is an R-flat proper overring of R. Thus each prime ideal of $\Theta(M)$ is extended from a prime ideal of R. Moreover, if Q is a nonzero prime ideal of $\Theta(M)$, then $P = Q \bigcap R$ is the (unique) prime ideal that extends to Q. If P is not comaximal with M, then we also have $PR_M \bigcap R = P$. Hence it must be that $P + M = R$ and therefore (vi) implies (v).

(vii) $\Rightarrow$ (ii) Let P be a nonzero prime ideal that is contained in M. If M is the only maximal ideal that contains P and P contains an invertible ideal J such that $\text{Max}(R, J) = \{M\}$, then $(R : J) \subseteq \Theta(M)$ since $(R : J) \subseteq R_N$ for each maximal ideal N that does not contain J. Hence $J\Theta(M) = \Theta(M)$ and thus we also have $P\Theta(M) = \Theta(M)$.

(i) $\Rightarrow$ (vii) Let P be a nonzero prime ideal that is contained in M. From above, we know $P\Theta(M) = \Theta(M)$. Thus M is the only maximal ideal that contains P. Let $r \in P\backslash\{0\}$ and set $A := rR_M \bigcap R$. Then each minimal prime of A is contained in M, and thus M is the only maximal ideal that contains A. By Theorem 6.1.3, $(R : A)R_M = (R_M : AR_M)$. Since $AR_M = rR_M$, AR_M is invertible in R_M and thus $A(R : A)R_M = R_M$. We also have $A\Theta(M) = \Theta(M)$. Hence $A \subseteq P$ is invertible with $\text{Max}(R, A) = \{M\}$. This completes the proof of the equivalence of (i)–(vii).

□

The following is a slight generalization of a statement in [6, Corollary 4.4].

Lemma 6.1.5. *Let V be a valuation domain whose maximal ideal M is principal. If F is a field contained in V/M such that $[V/M : F] = 2$, then the ring R formed by taking the pullback of F over M is a pseudo-valuation domain such that each nonzero ideal is divisorial.*

Proof. Since M is a principal ideal of V, each nonzero ideal of V is divisorial. Each ideal of V is also an ideal of R. By [30, Corollary 2.9], each nonzero ideal of V is a divisorial ideal of R. By [30, Remark 2.11], if J is an ideal of R that is not a principal ideal of R, then $J^v = JV$. So it suffices to show that each nonprincipal ideal of R is an ideal of V.

By way of contradiction, suppose A is a nonzero nonprincipal ideal of R that is not an ideal of V. Then there is an element $q \in A^v(= AV)\backslash A$ and, moreover, $A(R : A) \subseteq M$. There must be an element $p \in A$ such that $q/p \in V$. If $q/p \in M$, then we have $q = p(q/p) \in AM \subseteq A$, a contradiction. Hence q/p is a unit of V. Note that if there is an element $s \in A$ whose valuation (under the valuation associated to V) is strictly smaller than the value of q, then $q/s \in M$ again gives the contradiction that $q \in A$. Hence A^v is a principal ideal of V and $qV = pV = A^v = AV$. Note that if $r \in A^v$ is such that $rV \subsetneq pV$, then $r/p \in M$ and we have $r \in A$. Consider the ideal $qR + pR$. Both p and q are in M so this is a proper ideal of R. We have $q/p \in V\backslash R$ is a unit of V, so it is also the case that $p/q \in V\backslash R$. Since $[V/M : R/M] = [V/M : F] = 2$, $V/M = F + (q/p)F$. If there is an element $f \in A$ such that $fV = AV$ and p does not divide f in R, then f/p is such that $V/M = F + (f/p)F$. But then $q/p = a + b(f/p)$ some $a, b \in F$ and from this we would have $q = a'p + b'f \in A$ for some $a', b' \in R$, a contradiction. Hence $A = pR$ is principal and thus a divisorial ideal of R. □

Theorem 6.1.6. *Let T be a Prüfer domain with an invertible maximal ideal M such that $[T/M : F] = 2$ for some field $F \subsetneq T/M$. Also, let R be the pullback of F over M.*

(1) T has weak factorization if and only if R has weak factorization.
(2) T has strong factorization if and only if R has strong factorization.
(3) T has pseudo-Dedekind factorization if and only if R has pseudo-Dedekind factorization.

Proof. For (1), assume T has weak factorization. Then each nonzero prime that is contained in M is contained in no other maximal ideal (Lemma 4.2.3). Since T is Prüfer domain, these primes are linearly ordered. Also each such prime contains an invertible ideal J of T such that M is the only maximal ideal that contains J (Proposition 4.2.2). For such an ideal J, there are elements $f \in J$ and $q \in (T : J)$ such that $qfT + M = T$. Hence there is a $t \in T$ and $m \in M$ such that $tqf + m = 1$ which puts $tqf \in R\backslash M$. Let $B = fR + J^2$. Since $tqJ \subseteq T$, $tqJ^2 \subseteq J \subseteq R$. Hence $tq \in (R : B)$ with $tq \in R\backslash M$. As M is the only maximal ideal of R that contains J, B is an invertible ideal of R with $\text{Max}(R, B) = \{M\}$. Thus M is an h-local maximal ideal of R. As $[T/M : R/M] = 2$, each nonzero ideal of R_M is divisorial (Lemma 6.1.5) and therefore by Theorem 6.1.3, $IR_M \bigcap R$ is a divisorial ideal of R for each nonzero ideal $I \subseteq M$.

Let I be an arbitrary nonzero ideal of R and let $A := IR_M \bigcap R$. If I is not contained in M, then $A = R$. Otherwise, A is a proper divisorial ideal of R and M is the only maximal ideal of R that contains A. Moreover, $A\Theta(M) = \Theta(M)$. If M is the only maximal ideal that contains I, then $I = A$ is divisorial. Thus we may assume I is contained in at least one maximal ideal other than M. Hence $B := I\Theta(M) \bigcap R$ is a proper ideal of R. By Theorem 6.1.3, B is comaximal with both

M and A. Thus $AB = A \bigcap B$. In addition, $(AB)^v = (A \bigcap B)^v = A \bigcap B^v = AB^v$ since A is divisorial.

We also have $(B\Theta(M))^v = B^v\Theta(M)$ with $B^v = (R : (R : B))$ (but the same equality holds if instead $B^v = (T : (T : BT)))$. By Theorem 4.3.8, $\Theta(M)$ has weak factorization. Hence $B\Theta(M) = B^v N_1^{r_1} N_2^{r_2} \cdots N_m^{r_m}\Theta(M)$ for some noninvertible maximal ideals $N_1, N_2, \ldots, N_m$ of R. We have $B = B\Theta(M) \bigcap R$ and clearly $B^v N_1^{r_1} N_2^{r_2} \cdots N_m^{r_m}$ is comaximal with M. Hence $B = B^v N_1^{r_1} N_2^{r_2} \cdots N_m^{r_m}$. It follows that $I = AB^v N^{r_1} N_2^{r_2} \cdots N_m^{r_m} = I^v N_1^{r_1} N_2^{r_2} \cdots N_m^{r_m}$. Therefore R has weak factorization.

For the converse, assume R has weak factorization. Let P be a nonzero nonmaximal prime ideal of R. Then $P = P^v A$ where A is either equal to R or to a finite product of maximal ideals of R. The latter can occur only if $P = P^v$, so in either case $P = P^v$. Similarly, if Q is a proper P-primary ideal, then Q is divisorial as we have $QR_P = Q^v R_P$.

Next we show that M is an h-local maximal ideal of both R and T. The proof for T is similar to that used to establish Lemma 4.2.3(1). Let P be a nonzero prime ideal of T that is properly contained in M. Then P is also a prime ideal of R. We may further assume that P is branched. Thus $P = \sqrt{B}$ for some invertible ideal B of T by Theorem 2.3.12. As in the proof of Lemma 4.2.3(1), let $J := BT_M \cap T$. Then by Theorem 2.5.2(2), $J^v = J(P' : P') = B(P' : P')$ where P' is the largest prime ideal that is contained in all of the maximal ideals of T that contain P. In addition $J^{-1} = B^{-1}P'$ and $J \subsetneq J^v$ whenever more than one maximal ideal contains P. Since $\operatorname{Max}(T, P) = \operatorname{Max}(T, P')$, P' is a sharp prime of T and thus a maximal ideal of $(P' : P')$. In addition, P' is a prime ideal of R and it is the only maximal ideal of $(P' : P')$ that contains P. It follows that $J^v T_{P'} = BT_{P'} = JT_{P'}$ and we have $J^v = BT_{P'} \cap T$. By way of contradiction, assume $P' \neq M$. Then P' is a (nonmaximal) prime ideal of R, $(P' : P') \supsetneq T$ and $T_{P'} = R_{P'}$. It follows that $J^v = BR_{P'} \cap R$. Since $J^v = J(P' : P') \supsetneq J$, J^v cannot be an invertible ideal of T. Moreover, we also have $J^v T_{P'} = J^v T_M$. Hence $J^v(T : J^v) \subseteq M$ and from this we have $(T : J) = (T : J^v) = (M : J^v) \subseteq (R : J^v) \subseteq (R : J)$. It follows that $(T : J^v) = (R : J) = (R : J^v)$ and therefore J^v is also the divisorial closure of J in R. Thus in R, $J = J^v H$ where H is a finite product of maximal ideals of R.

By Lemma 2.5.1(2), $JT_{P'} = J^v\Gamma(P)$. Hence $BT_M = JT_M \supseteq J\Gamma(P) = J^v H\Gamma(P) = JHT_{P'} = JT_{P'} = BT_{P'} \supseteq BT_M$. Since B is an invertible ideal of T, we have a contradiction as we assumed $P' \neq M$. Hence $P' = M$ and therefore each nonzero prime ideal of T that is contained in M is both sharp and contained in no other maximal ideal. Thus M is an h-local maximal ideal of T (Theorem 6.1.4). Since R_M is a pseudo-valuation domain with integral closure T_M, each proper overring of R_M contains T_M. Hence it must be that $R_M\Theta(M) = K$ and therefore M is an h-local maximal ideal of R as well.

Reset notation and now simply let J be a nonzero ideal of R. By Theorem 6.1.3, $JR_M \cap R$ and $J\Theta(M) \cap R$ are comaximal ideals of R. Clearly, their intersection is J. Moreover, since each nonzero ideal of R_M is divisorial, $JR_M \cap R$ is a divisorial ideal of R (Theorem 6.1.3). The ideal $J\Theta(M) \cap R$ is also comaximal with M. Hence

$J\Theta(M) \cap R$ is a divisorial ideal of R if and only if $J\Theta(M)$ is a divisorial ideal of $\Theta(M)$. So as in the Prüfer case, $\Theta(M)$ inherits weak factorization in a natural way from R.

Finally for a nonzero ideal I of T, we can split I into comaximal factors $IT_M \cap T$ and $C := I\Theta(M) \cap T$. As with $JR_M \cap R$, $IT_M \cap T$ is a divisorial ideal of T. Also $C^v\Theta(M) = (I\Theta(M))^v$ with $C^v = (I\Theta(M))^v \cap T$. If C is divisorial, then so is I as it is the product of comaximal divisorial ideals. On the other hand, if C is not divisorial, $I\Theta(M) = C\Theta(M) = C^v N\Theta(M)$ where N is a finite product of maximal ideals of T (distinct from M). In this case we have $I = I^v N$ (with $I^v = (IT_M \cap T)(C^v\Theta(M) \cap T)$). Therefore T has weak factorization.

For the equivalence in (2), the factorizations obtained in the proof of (1) are inherited from the factorizations in $\Theta(M)$. Thus the following are equivalent: (i) T has strong factorization, (ii) $\Theta(M)$ has strong factorization, (iii) R has strong factorization.

The statement in (3) follows easily from Theorems 5.1.8 and 5.2.4 and Corollary 5.2.14. □

6.2 Independent Pairs of Overrings

Recall that if V and W are incomparable valuation domains with the same quotient field K, then K is also the quotient field of $V \bigcap W$ and $V \bigcap W$ is a Bézout domain with exactly two maximal ideals, one is the contraction of the maximal ideal of V and the other is the contraction of the maximal ideal of W (see, for example, [34, Theorem 22.8]). Such a pair of valuation domains is said to be *independent* if (0) is the only common prime ideal. Since each overring of V has the form V_P for some prime ideal $P = PV_P$ of V and each overring of W has the form W_Q for some prime ideal $Q = QW_Q$ of W, the following are equivalent for V and W.

(i) V and W are independent.
(ii) $VW = K$.
(iii) No nonzero prime ideal of $V \bigcap W$ survives in both V and W.

We extend the notion of independent valuation domains as follows. For a pair of domains S and T with the same quotient field K, we say that S and T are *independent* if $ST = K$ and no nonzero prime ideal of $S \bigcap T$ survives in both S and T. In the event $S \bigcap T$ also has quotient field K, then all that one needs to check is that no nonzero prime ideal of $S \bigcap T$ survives in both S and T. In fact, a slightly weaker condition suffices.

Lemma 6.2.1. *Let S and T be overrings of a domain R. If no nonzero prime ideal of R survives in both S and T, then S and T are independent.*

Proof. Let K be the quotient field of R and assume no nonzero prime ideal of R survives in both S and T. Next, let V be a valuation domain with quotient field K that contains ST and let M be the maximal ideal of V. Then $M \bigcap S$ is a prime ideal of S, $M \bigcap T$ is a prime ideal of T and $M \bigcap R$ is a prime ideal of R. Since both $M \bigcap S$ and $M \bigcap T$ contain $M \bigcap R$, it must be that $M \bigcap R = (0)$. It follows that $M = (0)$ (since R and V have the same quotient field) and therefore $V = K$. Also note that if Q is a nonzero prime ideal of $S \bigcap T$, then $Q \bigcap R$ is a nonzero prime ideal of R. Thus Q survives in at most one of S and T. Therefore S and T are independent. □

In general, just knowing that no nonzero prime ideal of $S \bigcap T$ survives in both S and T is not enough to conclude that $ST = K$. Consider the domains $S := F[X]$ and $T := F[1/X]$ where F is field (and X an indeterminate over F). Then $S \bigcap T = F$, so trivially no nonzero prime ideal of $S \bigcap T$ survives in both S and T. However, $ST = F[X, 1/X]$ is properly contained in its quotient field, $F(X)$, and thus S and T are not independent.

Theorem 6.2.2. *Let R be a domain with a pair of proper overrings S and T such that $R = S \bigcap T$. If S and T are independent, then*

(1) both S and T are R-flat, and
(2) $(R : I)S = (S : IS)$ and $(R : I)T = (T : IT)$ for each nonzero ideal I of R.

Proof. To see that both S and T are R-flat, let P be a nonzero prime ideal of R that survives in S. Since S and T are independent, $PT = T$. Moreover, if $Q \subseteq P$ is a nonzero prime ideal of R, then $QS \neq S$ and so we also have $QT = T$. Both R_P and T contain R and the only prime ideals of R that survive in R_P are those that are contained in P. Hence no nonzero prime ideal of R survives in both R_P and T. Thus R_P and T are independent by Lemma 6.2.1, and we have $T_P = TR_P = K$. Then $R_P = (S \bigcap T)_P = S_P \bigcap T_P = S_P$. That S is R-flat follows from [74, Theorem 1]. A similar proof shows that T is R-flat.

For (2), first note that since $S \bigcap T = R$, $(R : I) = (S : IS) \bigcap (T : IT)$. Hence $(R : I)S = [(S : IS) \bigcap (T : IT)]S$. Since S is R-flat, S distributes over the intersection and we have $(R : I)S = (S : IS) \bigcap (T : IT)S = (S : IS)$ since $ST = K$. Similarly, $(R : I)T = (T : IT)$. □

In general, for an independent pair S and T, a nonzero proper ideal I of $S \bigcap T$ can be such that $IS = S$ and $IT = T$. For example, let $R := K[X, Y]$ and let $S := R_{(X)}$ and $T := \bigcap\{R_{(f)} \mid f \in R\backslash(X) \text{ is irreducible}\}$. Clearly, S and T are independent with $S \bigcap T = R$, but the maximal ideal $XR + YR$ blows up in both S and T.

In the next result, we consider what additional conclusions one can draw when S and T are independent and each nonzero ideal of $S \bigcap T$ survives in at least one of S and T. It is helpful at this point to introduce the notion of "splitting sets" for $\mathrm{Max}(R)$.

Let $\mathcal{Y}$ be a nonempty subset of $\mathrm{Max}(R)$ and let $R_{\mathcal{Y}} := \bigcap\{R_M \mid M \in \mathcal{Y}\}$. We say that a pair of nonempty subsets $\mathcal{Y}_1$ and $\mathcal{Y}_2$ of $\mathrm{Max}(R)$ *split* R if the corresponding overrings $R_{\mathcal{Y}_1}$ and $R_{\mathcal{Y}_2}$ are independent with $R = R_{\mathcal{Y}_1} \bigcap R_{\mathcal{Y}_2}$ (so necessarily, $\mathcal{Y}_1$ and $\mathcal{Y}_2$ are disjoint). If, in addition, $\mathrm{Max}(R) = \mathcal{Y}_1 \bigcup \mathcal{Y}_2$, then we say that the pair $\mathcal{Y}_1$ and $\mathcal{Y}_2$ *fully split* R.

Theorem 6.2.3. *Let R be a domain with a pair of proper independent overrings S and T such that $R = S \bigcap T$ and each nonzero ideal of R survives in at least one of S and T.*

(1) Each nonzero prime ideal of R survives in exactly one of S and T.
(2) $\mathrm{Max}(S) = \{MS \mid M \in \mathrm{Max}(R), MS \subsetneq S\}$ and $\mathrm{Max}(T) = \{NT \mid N \in \mathrm{Max}(R), NT \subsetneq T\}$.
(3) The sets $\mathcal{M}_S := \{M \in \mathrm{Max}(R) \mid MS \neq S\}$ and $\mathcal{M}_T := \{N \in \mathrm{Max}(R) \mid NT \neq T\}$ fully split R. Moreover, $S = \bigcap\{R_M \mid MS \in \mathrm{Max}(S)\}$ and $T = \bigcap\{R_N \mid NT \in \mathrm{Max}(T)\}$.
(4) For each nonzero ideal I of R, the ideals $I_S := IS \bigcap R$ and $I_T := IT \bigcap R$ are comaximal with $I = I_S I_T$. Moreover, $I = IS \bigcap IT$, $I_S T = T$ and $I_T S = S$.
(5) For each nonzero ideal I of R, $I^v S = (IS)^v$, $(I_S)^v = (I^v)_S$, $I^v T = (IT)^v$ and $(I_T)^v = (I^v)_T$ (where $(IS)^v = (S : (S : IS))$ and $(IT)^v = (T : (T : IT))$).

Proof. By assumption, each nonzero prime ideal of R survives in at least one of S and T, and thus in exactly one since S and T are independent. In particular, if M is a maximal ideal of R, then either $MS \neq S$ with $MT = T$ or $MT \neq T$ with $MS = S$.

By flatness, if P is a nonzero prime ideal of R such that $PS \neq S$, then PS is a prime ideal of S such that $PS \bigcap R = P$. Moreover, each nonzero prime of S is extended from a nonzero prime ideal of R which blows up in T. Hence $\mathrm{Max}(S) = \{MS \mid M \in \mathrm{Max}(R)$ with $MS \neq S\}$ and $S = \bigcap\{R_M \mid MS \in \mathrm{Max}(S)\}$. Similarly $\mathrm{Max}(T) = \{NT \mid N \in \mathrm{Max}(R), NT \subsetneq T\}$ and $T = \bigcap\{R_N \mid NT \in \mathrm{Max}(T)\}$. Since each maximal ideal of R survives in exactly one of S and T, the sets $\mathcal{M}_S$ and $\mathcal{M}_T$ fully split R.

For (4) and (5), let I be a nonzero ideal of R. Since $\mathcal{M}_S$ and $\mathcal{M}_T$ fully split R with $S = R_{\mathcal{M}_S}$ and $T = R_{\mathcal{M}_T}$, then $I_S \bigcap I_T = I$. To see that I_S and I_T are comaximal, suppose P is a minimal prime of I_S. If $PS = S$, then there is a finitely generated ideal $B \subseteq P$ such that $PS = BS$. Since P is minimal over I_S, there is a positive integer n and an element $t \in R \backslash P$ such that $tB^n \subseteq I$. It follows that $tS = tB^n S \subseteq IS = I_S S$ which leads to the contradictory containment $t \in I_S \subseteq P$. Hence $PS \subsetneq S$ and $PT = T$. It follows that I_S and I_T are comaximal and therefore $I = I_S I_T$. Moreover, no prime ideal of T can contain $I_S T$ as the contraction of such a prime to R would contain a minimal prime of I_S. Hence $I_S T = T$. We also have $I_T S = S$. That $I = IS \bigcap IT$ follows from the above and the fact that $R = S \bigcap T$.

By Theorem 6.2.2, $(S : I^v S) = (R : I^v)S = (R : I)S = (S : IS)$ and $(T : I^v T) = (R : I^v)T = (R : I)T = (T : IT)$. So $(IS)^v = (I^v S)^v$ and $(IT)^v = (I^v T)^v$. Let $J_S = (IS)^v \bigcap R$ and $J_T = (IT)^v \bigcap R$. Then $J_S \supseteq (I^v)_S$ and $J_T \supseteq (I^v)_T$. It follows that $J := J_S J_T$ contains I^v. Since J_S contains I_S and J_T contains I_T, J_S and J_T are comaximal. Also, $J_S T = T$ and $J_T S = S$. With regard to duals we have $(R : I) \supseteq (R : J) = (S : JS) \bigcap (S : JT) = (S : IS) \bigcap (T : IT) = (R : I)$. Thus $J = I^v$ and therefore $I^v S = JS = J_S S = (IS)^v = (I^v S)^v$ and $I^v T = JT = J_T = (IT)^v = (I^v T)^v$. We also have $(I^v)_S = J_S$ and $(I^v)_T = J_T$. Since I_S and I_T are comaximal and $I = I_S I_T$, $I^v = (I_S)^v (I_T)^v$. In addition, $(I^v)_S = (I_S)^v$ and $(I^v)_T = (I_T)^v$. □

6.3 Jaffard Families and Matlis Partitions

Let $\mathcal{S} := \{S_\alpha\}_{\alpha\in\mathcal{A}}$ be a family of domains (that are not fields) with the same quotient field K such that $R := \bigcap_{\alpha\in\mathcal{A}} S_\alpha$ also has quotient field K. For each nonzero ideal I of R and each $\alpha \in \mathcal{A}$, let $I_\alpha := IS_\alpha \bigcap R$ and let $\text{supp}_{\mathcal{S}}(I) := \{\alpha \in \mathcal{A} \mid I_\alpha \neq R\}$ (= the *support of I with respect to $\mathcal{S}$*). It is clear that $\alpha \in \text{supp}_{\mathcal{S}}(I)$ if and only if $IS_\alpha \neq S_\alpha$. Also if $I \subseteq J$, then $\text{supp}_{\mathcal{S}}(I) \supseteq \text{supp}_{\mathcal{S}}(J)$. We say that such a family is a *Jaffard family* if for each nonzero ideal I of R,

(a) $\text{supp}_{\mathcal{S}}(I)$ is a finite nonempty subset of $\mathcal{A}$,
(b) $I = I_{\alpha_1} I_{\alpha_2} \cdots I_{\alpha_n}$ where $\text{supp}_{\mathcal{S}}(I) = \{\alpha_1, \alpha_2, \ldots, \alpha_n\}$, and
(c) $I_\alpha + I_\beta = R$ for all $\alpha \neq \beta$ in $\mathcal{A}$.

Clearly, it is enough to check (c) for $\alpha \neq \beta$ in $\text{supp}_{\mathcal{S}}(I)$. Also, since $R = \bigcap_{\alpha\in\mathcal{A}} S_\alpha$, (b) implies $I = \bigcap_{\alpha\in\mathcal{A}} IS_\alpha$ with $IS_\alpha = S_\alpha$ for all but finitely many α.

Theorem 6.3.1. *Let $\mathcal{S} := \{S_\alpha\}_{\alpha\in\mathcal{A}}$ be a Jaffard family with $R := \bigcap_{\alpha\in\mathcal{A}} S_\alpha$. Also for each $\alpha \in \mathcal{A}$, let $T_\alpha := \bigcap\{S_\gamma \mid \gamma \in \mathcal{A}\backslash\{\alpha\}\}$ (= K, if $|\mathcal{A}| = 1$).*

(1) For each nonzero prime ideal P of R, $|\text{supp}_{\mathcal{S}}(P)| = 1$ and $P = P_\beta$ when $PS_\beta \subsetneq S_\beta$. Moreover, if Q is a prime ideal of R such that $\text{supp}_{\mathcal{S}}(Q) = \{\beta\} = \text{supp}_{\mathcal{S}}(P)$, then $PS_\beta \subsetneq QS_\beta$ if and only if $P \subsetneq Q$.
(2) The following are equivalent for each pair of ideals A and B of R.

(i) $A = B$.
(ii) $A_\alpha = B_\alpha$ for each $\alpha \in \mathcal{A}$.
(iii) $AS_\alpha = BS_\alpha$ for each $\alpha \in \mathcal{A}$.
(iv) $\text{supp}_{\mathcal{S}}(A) = \text{supp}_{\mathcal{S}}(B)$ and $AS_\beta = BS_\beta$ for each $\beta \in \text{supp}_{\mathcal{S}}(A)$.

(3) For each nonzero ideal I of R and each pair $\alpha \neq \beta$ in $\mathcal{A}$, $I_\alpha S_\beta = S_\beta$.
(4) For each $\alpha \in \mathcal{A}$, S_α and T_α are independent, so both are R-flat.
(5) For each $\alpha \in \mathcal{A}$ and each maximal ideal N of S_α, $N \cap R$ is a maximal ideal of R and $N = (N \cap R)S_\alpha$.
(6) For each nonzero ideal I of R, I is invertible as an ideal of R if and only if IS_α is an invertible ideal of S_α for each α (equivalently, for each $\alpha \in \text{supp}_{\mathcal{S}}(I)$).
(7) For each nonzero ideal I of R and each $\alpha \in \mathcal{A}$, $(R : I)S_\alpha = (S_\alpha : IS_\alpha)$ and $(IS_\alpha)^v = (I^v S_\alpha)^v$.
(8) For each nonzero ideal I of R and each $\alpha \in \mathcal{A}$, $(I^v)_\alpha = (I_\alpha)^v$, $I^v S_\alpha = (IS_\alpha)^v$.

Proof. First for a nonzero prime P of R, let $\{\beta_1, \beta_2, \ldots, \beta_n\} = \text{supp}_{\mathcal{S}}(P)$. We have $P = P_{\beta_1} P_{\beta_2} \cdots P_{\beta_n}$ with $P_{\beta_i} + P_{\beta_j} = R$ for all $i \neq j$. Since P is a prime ideal and each P_{β_i} contains P, we have $n = 1$. Thus $P = P_{\beta_1}$ and $PS_\alpha = S_\alpha$ for all $\alpha \in \mathcal{A}\backslash\text{supp}_{\mathcal{S}}(P)$. Let Q be a nonzero prime ideal with $\text{supp}_{\mathcal{S}}(Q) = \{\beta_1\}$. We have $Q = Q_{\beta_1}$ and $P = P_{\beta_1}$, so clearly $P \subsetneq Q$ if and only if $PS_{\beta_1} \subsetneq QS_{\beta_1}$.

The statement in (2) is clear from the factorization property for nonzero ideals.

Let I be a nonzero ideal of R and let $\alpha \neq \beta$ be indices in $\mathcal{A}$. Then $(I_\alpha)_\beta = I_\alpha S_\beta \cap R \supseteq IS_\beta \cap R = I_\beta$ also $(I_\alpha)_\beta \supseteq I_\alpha$. As I_α and I_β are comaximal, we have $I_\alpha S_\beta = S_\beta$.

For (4), it is clear that $R = S_\beta \bigcap T_\beta$ for each $\beta \in \mathscr{A}$. Since $S_\beta \subsetneq K$, there is at least one nonzero prime ideal P such that $PS_\beta \neq S_\beta$, necessarily with $P = P_\beta$. We will show that $PT_\beta = T_\beta$ for each such prime P. Let $r \in P$ be nonzero. Then rR factors as $rR = (rR)_\beta (rR)_{\alpha_1} (rR)_{\alpha_2} \cdots (rR)_{\alpha_n}$ with $\mathrm{supp}_{\mathscr{S}}(rR) = \{\beta, \alpha_1, \alpha_2, \ldots, \alpha_n\}$ (possibly with no α_is). Clearly, the ideal $(rR)_\beta$ is invertible. We also have $(rR)_\beta S_\alpha = S_\alpha$ for each $\alpha \neq \beta$. Thus each such S_α contains $(R : (rR)_\beta)$ and therefore $(R : (rR)_\beta) \subseteq T_\beta$. Hence $(rR)_\beta T_\beta = T_\beta$. As $P = P_\beta \supseteq (rR)_\beta$, $PT_\beta = T_\beta$. It follows that no nonzero prime ideal of R survives in both S_β and T_β. Therefore S_β and T_β are independent by Lemma 6.2.1. Hence each is R-flat by Theorem 6.2.2.

For (5), S_β is flat over R so each prime ideal of S_β is extended from a prime ideal of R. In particular, if N is a maximal ideal of S_β, then $P := N \bigcap R$ is a prime ideal of R. By (1), each maximal ideal M of R that contains P is such that $\mathrm{supp}_{\mathscr{S}}(M) = \{\beta\}$. Thus we have $N = PS_\beta \subseteq MS_\beta \subsetneq S_\beta$. Hence $P = M$ and $N = MS_\beta$.

Clearly, if I is an invertible ideal of R, then IS_α is an invertible ideal of S_α for each $\alpha \in \mathscr{A}$. To establish the converse, first note that $IS_\alpha = S_\alpha$ except for those $\alpha_i \in \mathrm{supp}(I)$, a finite set. Thus we start with the assumption that IS_{α_i} is an invertible ideal for each $\alpha_i \in \mathrm{supp}_{\mathscr{S}}(I) = \{\alpha_1, \alpha_2, \ldots, \alpha_n\}$. Hence there is a finitely generated ideal $J \subseteq I$ such that $JS_{\alpha_i} = IS_{\alpha_i}$ for each $1 \leq i \leq n$. While it may be that $JS_\alpha \neq S_\alpha$ for some $\alpha \in \mathscr{A} \backslash \mathrm{supp}_{\mathscr{S}}(I)$, there are only finitely many such α. Thus, in this event, there is a finitely generated ideal $B \subseteq I$ such that $BS_\alpha = S_\alpha$ for each $\alpha \in \mathrm{supp}_{\mathscr{S}}(J) \backslash \mathrm{supp}_{\mathscr{S}}(I)$. The finitely generated ideal $J + B$ is contained in I with $\mathrm{supp}_{\mathscr{S}}(J+B) = \mathrm{supp}_{\mathscr{S}}(I)$ and $(J+B)S_{\alpha_i} = IS_{\alpha_i}$ for each $\alpha_i \in \mathrm{supp}_{\mathscr{S}}(I)$ and $(J + B)S_\beta = S_\beta = IS_\beta$ for each $\beta \in \mathscr{A} \backslash \mathrm{supp}_{\mathscr{S}}(I)$. Hence $I = J + B$ and we at least have that I is a finitely generated of R. To complete the proof, it suffices to show IR_M is invertible (principal) for each maximal ideal M of R. Let M be a maximal ideal that contains I. Then $\mathrm{supp}_{\mathscr{S}}(M) = \{\alpha_i\}$ for some $1 \leq i \leq n$ since each maximal ideal survives in exactly one S_α and certainly $MS_\alpha = S_\alpha$ whenever $IS_\alpha = S_\alpha$. By flatness $R_M = (S_{\alpha_i})_M$ and thus $IR_M = (IS_{\alpha_i})_M$ is an invertible ideal of R_M.

For (7) and (8), let I be a nonzero ideal of R and let $\alpha \in \mathscr{A}$. It is clear that each nonzero ideal of R survives in at least one S_α and T_α. Thus we may apply Theorems 6.2.2 and 6.2.3 to obtain all of the conclusions in (7) and (8): $(R : I)S_\alpha = (S_\alpha : IS_\alpha)$, $(IS_\alpha)^v = (I^v S_\alpha)^v$, $(I^v)_\alpha = (I_\alpha)^v$ and $I^v S_\alpha = (IS_\alpha)^v$. □

For a domain R, let $\mathrm{Spec}^*(R) := \mathrm{Spec}(R) \backslash \{(0)\}$.

Corollary 6.3.2. *Let $\mathscr{S} := \{S_\alpha\}_{\alpha \in \mathscr{A}}$ be a Jaffard family and let $R := \bigcap S_\alpha$. Also, for each $\alpha \in \mathscr{A}$, let $\mathscr{X}_\alpha := \{M \in \mathrm{Max}(R) \mid MS_\alpha \neq S_\alpha\}$ and $\mathscr{Y}_\alpha := \{P \in \mathrm{Spec}^*(R) \mid PS_\alpha \neq S_\alpha\}$. Then the collection of sets $\{\mathscr{X}_\alpha\}_{\alpha \in \mathscr{A}}$ partitions $\mathrm{Max}(R)$, and for each α, $\mathrm{Max}(S_\alpha) = \{MS_\alpha \mid M \in \mathscr{X}_\alpha\}$ and $S_\alpha = \bigcap \{R_M \mid M \in \mathscr{X}_\alpha\}$. Also, the collection of sets $\{\mathscr{Y}_\alpha\}_{\alpha \in \mathscr{A}}$ partitions $\mathrm{Spec}^*(R)$, and for each α, $\mathrm{Spec}^*(S_\alpha) = \{PS_\alpha \mid P \in \mathscr{Y}_\alpha\}$.*

Proof. By Theorem 6.3.1(1), if P is a nonzero prime ideal of R, then there is a unique S_α such that $PS_\alpha \neq S_\alpha$. So necessarily, if M is a maximal ideal of R that contains P, then $PS_\alpha \neq S_\alpha$ implies $MS_\alpha \neq S_\alpha$. Thus it is clear that $\{\mathscr{X}_\alpha\}_{\alpha\in\mathscr{A}}$ partitions $\mathrm{Max}(R)$ and $\{\mathscr{Y}_\alpha\}_{\alpha\in\mathscr{A}}$ partitions $\mathrm{Spec}^*(R)$. Each S_α is R-flat by Theorem 6.3.1(4). Thus if Q is a prime ideal of S_α, then $Q = (Q \cap R)S_\alpha$. Also, if P is a prime ideal of R such that $PS_\alpha \neq S_\alpha$, then $PS_\alpha \cap R = P$ with PS_α a prime ideal of S_α. Thus $\mathrm{Max}(S_\alpha) = \{MS_\alpha \mid M \in \mathscr{X}_\alpha\}$ and $\mathrm{Spec}^*(S_\alpha) = \{PS_\alpha \mid P \in \mathscr{Y}_\alpha\}$ for each α.

For each $M \in \mathscr{X}_\alpha$, $R_M = (S_\alpha)_{MS_\alpha}$ by flatness. Thus $S_\alpha = \bigcap\{R_M \mid M \in \mathscr{X}_\alpha\}$. □

Next, we take a different approach which generalizes the notion of splitting pairs. Let $\mathscr{P} := \{\mathrm{X}_\alpha\}_{\alpha\in\mathscr{A}}$ be a partition of $\mathrm{Max}(R)$ and for each $\alpha \in \mathscr{A}$, let $W_\alpha := \bigcap\{R_M \mid M \in \mathrm{X}_\alpha\}$. As above, let $\mathrm{supp}_{\mathscr{P}}(I) = \{\alpha \in \mathscr{A} \mid IW_\alpha \subsetneq W_\alpha\}$(=the *support of I with respect to $\mathscr{P}$*. Say that $\mathscr{P}$ is a *Matlis partition* of $\mathrm{Max}(R)$ if $|\mathrm{supp}_{\mathscr{P}}(rR)| < \infty$ for each nonzero nonunit $r \in R$ and $|\mathrm{supp}_{\mathscr{P}}(P)| = 1$ for each nonzero prime ideal P of R. Note that $\mathrm{supp}_{\mathscr{P}}(I)$ is nonempty (but finite) for each nonzero ideal I of R. Also, R is h-local if and only if $\mathscr{P} := \{\{M_\alpha\} \mid M_\alpha \in \mathrm{Max}(R)\}$ is a Matlis partition of R.

Lemma 6.3.3. *Let R be an integral domain and let $\mathscr{P} := \{\mathrm{X}_\alpha\}_{\alpha\in\mathscr{A}}$ be a Matlis partition of* $\mathrm{Max}(R)$. *For each nonzero ideal I of R and each $\beta \in \mathrm{supp}_{\mathscr{P}}(I)$, if P is a minimal prime ideal of $IW_\beta \cap R$, then $\mathrm{supp}_{\mathscr{P}}(P) = \{\beta\}$.*

Proof. Note that for the ideal I, $I_\beta := IW_\beta \cap R$ is such that $I_\beta W_\beta = IW_\beta$. So we may assume $I = I_\beta$. Let P be a minimal prime ideal of I and by way of contradiction assume $\mathrm{supp}_{\mathscr{P}}(P) \neq \{\beta\}$. Since $|\mathrm{supp}_{\mathscr{P}}(P)| = 1$, it must be that $PW_\beta = W_\beta$. Thus there is a finitely generated ideal $B \subseteq P$ such that $BW_\beta = W_\beta$. Since P is minimal over I (and B is finitely generated), some power of BR_P is contained in IR_P. Without loss of generality, we may assume $BR_P \subseteq IR_P$. Thus there is an element $t \in R\backslash P$ such that $tB \subseteq I$. Extending to W_β, we get $tW_\beta = tBW_\beta \subseteq IW_\beta$ and from this we get $t \in IW_\beta \cap R = I$, a contradiction. Hence $\mathrm{supp}_{\mathscr{P}}(P) = \{\beta\}$. □

Theorem 6.3.4. *Let R be an integral domain with quotient field $K \supsetneq R$.*

(1) *If $\mathscr{S} := \{S_\alpha\}_{\alpha\in\mathscr{A}}$ is a Jaffard family such that $R = \bigcap S_\alpha$, then there is a Matlis partition $\mathscr{P} := \{\mathrm{X}_\alpha\}_{\alpha\in\mathscr{A}}$ of* $\mathrm{Max}(R)$ *such that $S_\alpha = W_\alpha$ for each $\alpha \in \mathscr{A}$.*
(2) *If $\mathscr{P} := \{\mathrm{X}_\alpha\}_{\alpha\in\mathscr{A}}$ is a Matlis partition of* $\mathrm{Max}(R)$, *then $\{W_\alpha\}_{\alpha\in\mathscr{A}}$ is a Jaffard family such that $R = \bigcap W_\alpha$.*

Proof. The statement in (1) follows directly from Corollary 6.3.2.

For the proof of (2), suppose $\mathscr{P} := \{\mathrm{X}_\alpha\}_{\alpha\in\mathscr{A}}$ is a Matlis partition of $\mathrm{Max}(R)$. By definition each nonzero prime ideal of R survives in exactly one W_α. Thus by Lemma 6.2.1, W_α and W_β are independent for each pair $\alpha \neq \beta$ in $\mathscr{A}$.

Next let I be a nonzero ideal of R. Then $\mathrm{supp}_{\mathscr{P}}(I)$ is a finite set, say $\mathrm{supp}_{\mathscr{P}}(I) = \{\alpha_1, \alpha_2, \ldots, \alpha_n\}$. For each α_i, let $I_{\alpha_i} := IW_{\alpha_i} \cap R$. For $\alpha \in \mathscr{A}\backslash\mathrm{supp}_{\mathscr{P}}(I)$, $IW_\alpha = W_\alpha$ and so $I_\alpha := IW_\alpha \cap R = R$. We have

$I \subseteq \bigcap I_\alpha = \bigcap I_{\alpha_i}$ since $I_\alpha = R$ for all $\alpha \in \mathscr{A} \backslash \mathrm{supp}_{\mathscr{P}}(I)$. Also, for each $N \in \mathrm{X}_{\alpha_i}$, we have $W_{\alpha_i} \subseteq R_N$ and thus $IW_{\alpha_i} = I_{\alpha_i} W_{\alpha_i} \subseteq I_{\alpha_i} R_N = IR_N$. It follows that $I = \bigcap I_{\alpha_i}$.

By Lemma 6.3.3, $\mathrm{supp}_{\mathscr{P}}(P) = \{\alpha_i\}$ for each minimal prime P of I_{α_i}. It follows that $\mathrm{supp}_{\mathscr{P}}(M) = \{\alpha_i\}$ for each maximal ideal that contains I_{α_i}. Thus no maximal ideal can contain more than one I_{α_j} which means the I_{α_i}s are pairwise comaximal. Hence $I = \prod_{i=1}^n I_{\alpha_i}$. From this we have that $\{W_\alpha\}_{\alpha \in \mathscr{A}}$ is a Jaffard family (for R).

□

Theorem 6.3.5. *If* $\mathscr{S} := \{S_\alpha\}_{\alpha \in \mathscr{A}}$ *is a family of overrings of a domain R such that* $R = \bigcap_{\alpha \in \mathscr{A}} S_\alpha$, *then the following are equivalent.*

(i) $\mathscr{S}$ *is a Jaffard family.*
(ii) For each nonzero ideal I of R, $1 \leq |\mathrm{supp}_{\mathscr{S}}(I)| < \infty$ *and for each* $\beta \in \mathrm{supp}_{\mathscr{S}}(I)$, $\mathrm{supp}_{\mathscr{S}}(I_\beta) = \{\beta\}$.
(iii) (a) each nonzero nonunit of R is a unit in all but finitely many S_α*s,*
(b) for each nonzero ideal I of R, $IS_\alpha \neq S_\alpha$ *for at least one* $S_\alpha \in \mathscr{S}$*, and*
(c) for each $\beta \in \mathscr{A}$, S_β *and* $\bigcap_{\alpha \neq \beta} S_\alpha$ *are independent.*

Proof. By definition, each Jaffard family satisfies (a) and (b) in (iii), and each such family satisfies (ii) and (c) by Theorem 6.3.1.

To see that (iii) implies (ii), first note that (a) and (b) together imply $1 \leq |\mathrm{supp}_{\mathscr{S}}(I)| < \infty$ for each nonzero ideal I of R. Also, for each $\beta \in \mathrm{supp}_{\mathscr{S}}(I)$, the ideal $I_\beta (= IS_\beta \cap R)$ blows up in $\bigcap_{\alpha \neq \beta} S_\alpha$ by (c) and Theorem 6.2.3. It follows that $I_\beta S_\alpha = S_\alpha$ for each $\alpha \neq \beta$. Hence $\mathrm{supp}_{\mathscr{S}}(I_\beta) = \{\beta\}$. Therefore (iii) implies (ii).

To complete the proof, we show (ii) implies (i). Assume that for each nonzero ideal I of R, $1 \leq |\mathrm{supp}_{\mathscr{S}}(I)| < \infty$ and $\mathrm{supp}_{\mathscr{S}}(I_\beta) = \{\beta\}$ for each $\beta \in \mathrm{supp}_{\mathscr{S}}(I)$. If M is a maximal ideal of R, then $M_\beta = M$ for each $\beta \in \mathrm{supp}_{\mathscr{S}}(M)$ and thus $|\mathrm{supp}_{\mathscr{S}}(M)| = 1$.

First we show that for each nonzero ideal I, the ideals $I_{\alpha_1}, I_{\alpha_2}, \ldots, I_{\alpha_n}$ are pairwise comaximal and $\bigcap_{\alpha \in \mathscr{A}} IS_\alpha = \prod_{i=1}^n I_{\alpha_i}$ where $\mathrm{supp}_{\mathscr{S}}(I) = \{\alpha_1, \alpha_2, \ldots, \alpha_n\}$. That $I = \prod_{i=1}^n I_{\alpha_i}$ will then follow from showing that $S_\alpha = \bigcap\{R_M \mid M \in \mathrm{Max}(R)$ such that $MS_\alpha \neq S_\alpha\}$.

Let I be a nonzero ideal of R and let $\mathrm{supp}_{\mathscr{S}}(I) = \{\alpha_1, \alpha_2, \ldots, \alpha_n\}$. Since $R = \bigcap_{\alpha \in \mathscr{A}} S_\alpha$ and $IS_\alpha = S_\alpha$ for all $\alpha \notin \mathrm{supp}_{\mathscr{S}}(I)$, $I \subseteq \bigcap_{\alpha \in \mathscr{A}} IS_\alpha = I_{\alpha_1} \cap I_{\alpha_2} \cap \cdots \cap I_{\alpha_n}$. If M is a maximal ideal that contains I, then M survives in exactly one S_α which must be one of the S_{α_i}s. Clearly M contains the corresponding I_{α_i} but contains no other I_{α_j} (since $I_{\alpha_j} S_{\alpha_i} = S_{\alpha_i} \supsetneq MS_{\alpha_i}$ for all $j \neq i$). Hence the ideals $I_{\alpha_1}, I_{\alpha_2}, \ldots, I_{\alpha_n}$ are pairwise comaximal. Thus $\bigcap_{\alpha \in \mathscr{A}} IS_\alpha = \prod_{i=1}^n I_{\alpha_i}$. Since $I \subseteq \prod_{i=1}^n I_{\alpha_i}$ and $\mathrm{supp}_{\mathscr{S}}(I_{\alpha_i}) = \{\alpha_i\}$ for each i, $\mathrm{supp}_{\mathscr{S}}(\prod_{i=1}^n I_{\alpha_i}) = \mathrm{supp}_{\mathscr{S}}(I)$. Moreover, if $\{\gamma_1, \gamma_2, \ldots, \gamma_k\}$ is a nonempty subset of $\mathrm{supp}_{\mathscr{S}}(I)$, then $\mathrm{supp}_{\mathscr{S}}(\prod_{i=1}^k I_{\gamma_i}) = \{\gamma_1, \gamma_2, \ldots, \gamma_k\}$.

Next we show that there is an invertible ideal $B \subseteq I$ such that $\mathrm{supp}_{\mathscr{S}}(B) = \mathrm{supp}_{\mathscr{S}}(I)$. To start, let b be a nonzero element of I. Then $\mathrm{supp}_{\mathscr{S}}(I) \subseteq \mathrm{supp}_{\mathscr{S}}(bR) := \{\beta_1, \beta_2, \ldots, \beta_m\}$. We may assume $\beta_i = \alpha_i$ for $1 \leq i \leq n$. We are done if $m = n$, so we may further assume $m > n$. Since $R = \bigcap_{\alpha \in \mathscr{A}} S_\alpha$,

we have $bR = \bigcap_{\alpha\in\mathscr{A}} bS_\alpha$. Thus $bR = (\prod_{i=1}^{n}(bR)_{\alpha_i}) \cdot (\prod_{j=n+1}^{m}(bR)_{\beta_j})$. The ideals $B := \prod_{i=1}^{n}(bR)_{\alpha_i}$ and $C := \prod_{j=n+1}^{m}(bR)_{\beta_j}$ are comaximal invertible ideals with $bR = BC \subseteq I$. Since $\mathrm{supp}_{\mathscr{S}}(C) = \{\beta_{n+1}, \beta_{n+2}, \ldots, \beta_m\}$ has empty intersection with $\mathrm{supp}_{\mathscr{S}}(I)$, I and C are comaximal. Thus $B \subseteq I$ with $\mathrm{supp}_{\mathscr{S}}(B) = \mathrm{supp}_{\mathscr{S}}(I)$.

Let P be a nonzero prime ideal of R. Then from the argument in the previous paragraph, for each nonzero $d \in P$, $dR = (dR)_{\delta_1}(dR)_{\delta_2}\cdots(dR)_{\delta_m}$ where $\mathrm{supp}_{\mathscr{S}}(dR) := \{\delta_1, \delta_2, \ldots, \delta_m\}$ with the invertible ideals $(dR)_{\delta_1}, (dR)_{\delta_2}, \ldots, (dR)_{\delta_m}$ pairwise comaximal. It follows that P contains exactly one of the $(dR)_{\delta_j}$s and from this we deduce that $|\mathrm{supp}_{\mathscr{S}}(P)| = 1$.

Next we shift the focus to the S_αs. For a given S_β, there is an element $t \in R$ such that tS_β is a proper ideal of S_β (since each S_α is an overring of R and none are the quotient field). Thus the ideal $B_\beta := tS_\beta \cap R$ is an invertible ideal of R such that $\mathrm{supp}_{\mathscr{S}}(P) = \{\beta\}$ for each prime ideal P of R that contains B_β. Also, by assumption, $B_\beta S_\alpha = S_\alpha$ for all $\alpha \neq \beta$. It follows that each such S_α contains $(R : B_\beta)$ and thus so does $T_\beta := \bigcap_{\alpha\neq\beta} S_\alpha$. Moreover, if Q is a prime ideal of R such that $\mathrm{supp}_{\mathscr{S}}(Q) = \{\beta\}$, then Q contains an invertible ideal C such that $\mathrm{supp}_{\mathscr{S}}(C) = \{\beta\}$ and $CS_\beta \cap R = C$. As with B_β, $(R : C) \subseteq T_\beta$ and thus $QT_\beta = T_\beta$. Therefore $S_\beta T_\beta = K$, the quotient field of R. It follows that S_β and T_β are independent with $S_\beta \cap T_\beta = R$. By Theorem 6.2.3, $S_\beta = \bigcap\{R_M \mid M \in \mathrm{Max}(R)$ such that $MS_\beta \neq S_\beta\}$. □

Corollary 6.3.6. *Let $\mathscr{S} := \{S_\alpha\}_{\alpha\in\mathscr{A}}$ be a Jaffard family. If $S := \bigcap_{\alpha\in\mathscr{A}} S_\alpha$ is a Bézout domain, then for each nonzero ideal I, there is a principal ideal $bS \subseteq I$ such that* $\mathrm{supp}_{\mathscr{S}}(bS) = \mathrm{supp}_{\mathscr{S}}(I)$.

Proof. Let I be a nonzero ideal of S. Then from the proof of Theorem 6.3.5, there is an invertible ideal $B \subseteq I$ such that $\mathrm{supp}_{\mathscr{S}}(B) = \mathrm{supp}_{\mathscr{S}}(I)$. In the case S is Bézout, B is principal. □

Theorem 6.3.7. *Let $\mathscr{S} := \{S_\alpha\}_{\alpha\in\mathscr{A}}$ and be a Jaffard family with common quotient field K and let $\mathscr{B}$ be a nonempty subset of $\mathscr{A}$ with complement $\mathscr{C} := \mathscr{A}\backslash\mathscr{B}$.*

(1) If $\mathscr{B}$ is a proper subset of $\mathscr{A}$, then the domains $R_{\mathscr{B}} := \bigcap_{\beta\in\mathscr{B}} S_\beta$ and $R_{\mathscr{C}} := \bigcap_{\gamma\in\mathscr{C}} S_\gamma$ are independent.
(2) The set $\{S_\beta\}_{\beta\in\mathscr{B}}$ is a Jaffard family.
(3) If $\{W_\beta\}_{\beta\in\mathscr{B}}$ is a family of domains such that $S_\beta \subseteq W_\beta \subsetneq K$ for each $\beta \in \mathscr{B}$, then $\{W_\beta\}_{\beta\in\mathscr{B}}$ is a Jaffard family if and only if each nonzero ideal of $W := \bigcap_{\beta\in\mathscr{B}} W_\beta$ survives in at least one W_β.

Proof. Throughout the proof we let $R := \bigcap_{\alpha\in\mathscr{A}} S_\alpha$.

For (1), suppose $\mathscr{B}$ is a proper subset of $\mathscr{A}$ and let P be a nonzero prime ideal of R. Then by Theorem 6.3.1, $\mathrm{supp}_{\mathscr{S}}(P) = \{\tau\}$ for some $\tau \in \mathscr{A}$ and $PT_\tau = T_\tau$ where $T_\tau := \bigcap_{\alpha\neq\tau} S_\alpha$. It follows that $PR_{\mathscr{B}} = R_{\mathscr{B}}$ when $\tau \in \mathscr{C}$, and $PR_{\mathscr{C}} = R_{\mathscr{C}}$ when $\tau \in \mathscr{B}$. Therefore $R_{\mathscr{B}}$ and $R_{\mathscr{C}}$ are independent by Lemma 6.2.1.

For (2), there is nothing to prove if $\mathscr{B} = \mathscr{A}$, so we may assume $\mathscr{B}$ is a proper (nonempty) subset of $\mathscr{A}$. Let J be a nonzero ideal of $R_{\mathscr{B}}$ and let $I := J \cap R$. Since

$R_{\mathscr{B}}$ and $R_{\mathscr{C}}$ are independent and each nonzero ideal of R survives in at least one of these two domains, $IR_{\mathscr{C}} = R_{\mathscr{C}}$ and $IR_{\mathscr{B}} = J$ by Theorem 6.2.3. It follows that, for all $\alpha \in \mathscr{A}$, $JS_\alpha \neq S_\alpha$ if and only if $IS_\alpha \neq S_\alpha$. In addition, $JS_\gamma = S_\gamma$ for all $\gamma \in \mathscr{C}$. Thus $\mathrm{supp}_{\mathscr{S}}(I) \subseteq \mathscr{B}$ with $S_\beta \supsetneq IS_\beta = JS_\beta$ for each $\beta \in \mathrm{supp}_{\mathscr{S}}(I)$. Hence (with abuse of notation) $1 \leq |\mathrm{supp}_{\mathscr{S}}(J)| < \infty$.

For each $\beta \in \mathrm{supp}_{\mathscr{S}}(J)$, let $J_\beta := JS_\beta \bigcap R_{\mathscr{B}}$. Then $J_\beta \bigcap R = IS_\beta \bigcap R = I_\beta$. For $\delta \neq \beta$, $S_\delta \supseteq J_\beta S_\delta \supseteq I_\beta S_\delta = S_\delta$. Therefore $\{S_\beta\}_{\beta \in \mathscr{B}}$ is a Jaffard family by Theorem 6.3.5.

For (3), assume $S_\beta \subseteq W_\beta \subsetneq K$ for each $\beta \in \mathscr{B}$. Then $R \subseteq R_{\mathscr{B}} \subseteq W$ so each nonzero ideal of W contracts to a nonzero ideal of R. By definition, if $\{W_\beta\}_{\beta \in \mathscr{B}}$ is a Jaffard family, then $1 \leq \mathrm{supp}_{\mathscr{S}}(B) < \infty$ for each nonzero ideal B of W.

For the converse, assume each nonzero ideal of W survives in at least one W_β and let J be a nonzero ideal of W. Then $I := J \bigcap R$ is a nonzero ideal of R. Since $\mathscr{S}$ is a Jaffard family, I survives in at most finitely many S_αs. Clearly $IW_\beta = W_\beta$ whenever $IS_\beta = S_\beta$. Thus J survives in at most finitely many W_βs. Also $J_\beta := JW_\beta \bigcap W$ contains I_β. Thus for $\delta \in \mathscr{B} \backslash \{\beta\}$, $S_\delta = I_\beta S_\delta \subseteq J_\beta W_\delta$. It follows that $J_\beta W_\delta = W_\delta$. Therefore $\{W_\beta\}_{\beta \in \mathscr{B}}$ is a Jaffard family by Theorem 6.3.5. □

Theorem 6.3.8. *Let $\mathscr{S} := \{S_\alpha\}_{\alpha \in \mathscr{A}}$ be a Jaffard family with $R := \bigcap S_\alpha$.*

(1) Each S_α is integrally closed if and only if R is integrally closed.
(2) Each S_α is h-local if and only if R is h-local.
(3) Each S_α is a Prüfer domain if and only if R is a Prüfer domain.
(4) Each S_α has weak (strong) factorization if and only if R has weak (strong) factorization.
(5) Each S_α has (strong) pseudo-Dedekind factorization if and only if R has (strong) pseudo-Dedekind factorization.

Proof. A flat overring of an integrally closed domain is integrally closed, and an intersection of integrally closed domains is integrally closed. Also, by Theorem 6.3.1(5), each S_α is flat over R. Thus R is integrally closed if and only if each S_α is integrally closed.

By definition, each nonzero nonunit of R is a unit in all but finitely many S_αs. Hence by Corollary 6.3.2 it is easy to see that R is h-local if and only if each S_α is h-local.

An overring of a Prüfer domain is always a Prüfer domain, so if R is a Prüfer domain, then so is each S_α. By way of Theorem 6.3.1, there are several ways to establish the converse. For example by statement (6), a nonzero finitely generated ideal I of R is invertible if IS_α is an invertible ideal of S_α for each α. So clearly, R is Prüfer if each S_α is Prüfer.

Suppose each S_α has weak (strong) factorization and let I be a nonzero nondivisorial ideal of R with $\mathrm{supp}_{\mathscr{S}}(I) = \{\alpha_1, \alpha_2, \ldots, \alpha_n\}$. By Theorem 6.3.1 (7) & (8), at least one of $IS_{\alpha_1}, IS_{\alpha_2}, \ldots, IS_{\alpha_n}$ is not divisorial. Without loss of generality, we may assume IS_{α_i} is not divisorial for each $1 \leq i \leq m$ and IS_{α_j} is divisorial for each $m < j \leq n$. Then $I_{\alpha_j} S_{\alpha_j} = IS_{\alpha_j} = (I^v S_{\alpha_j})^v = (I^v)_{\alpha_j} S_{\alpha_j}$ for each $m < j \leq n$. And for $1 \leq i \leq m$, we have $I_{\alpha_i} S_{\alpha_i} = IS_{\alpha_i} = (IS_{\alpha_i})^v M_{i,1}^{r_{i,1}} M_{i,2}^{r_{i,2}} \cdots M_{i,s_i}^{r_{i,s_i}} S_{\alpha_i} =$

$(I^v)_{\alpha_i} M_{i,1}^{r_{i,1}} M_{i,2}^{r_{i,2}} \cdots M_{i,s_i}^{r_{i,s_i}} S_{\alpha_i}$ for some maximal ideals $M_{i,1}, M_{i,2}, \ldots, M_{i,s_i}$ ($s_i \geq 1$) which survive in S_{α_i} and positive integers $r_{i,1}, r_{i,2}, \ldots, r_{i,s_i}$. It follows that $I_{\alpha_j} = (I^v)_{\alpha_j}$ for $m < j \leq n$ and $I_{\alpha_i} = (I_{\alpha_i})^v M_{i,1}^{r_{i,1}} M_{i,2}^{r_{i,2}} \cdots M_{i,s_i}^{r_{i,s_i}}$ for $1 \leq i \leq m$. Since $(I_{\alpha_i})^v = (I^v)_{\alpha_i}$, we have $I = I^v \prod M_{i,k_i}^{r_{i,k_i}}$. Thus R has weak factorization.

Since each $M_{i,k}$ blows up in all S_αs other than S_{α_i}, if each $r_{i,k} = 1$, then the factorization for I has the form $I = I^v N_1 N_2 \cdots N_t$ with the N_ks distinct maximal ideals. Thus R has strong factorization if each S_α has strong factorization.

For the converse, suppose R has weak (strong) factorization and let J be a nonzero nondivisorial ideal of S_β for some β. By flatness, $I := J \bigcap R$ is such that $IS_\beta = J$. By Theorem 6.3.1(9), we know $I^v S_\beta = (IS_\beta)^v = J^v$ (with J^v the divisorial closure of J with respect to S_β). Thus I is not a divisorial ideal of R. Hence we may factor it as $I = I^v M_1^{r_1} M_2^{r_2} \cdots M_n^{r_n}$ for some maximal ideals $M_1, M_2, \ldots, M_n$ of R. Since $I = I_\beta$, $IS_\alpha = S_\alpha$ for each $\alpha \neq \beta$. As each M_i survives in a unique S_α, it must be that each extends to a maximal ideal of S_β (another consequence of flatness). Thus $J = I^v M_1^{r_1} M_2^{r_2} \cdots M_n^{r_n} S_\beta$ is a weak factorization for J in S_β. Therefore S_β has weak factorization. As above, if each $r_i = 1$, then we have a "strong" factorization of J in S_β. So S_β has strong factorization whenever R does.

Pseudo-Dedekind factorization is much easier since we know a domain T has pseudo-Dedekind factorization if and only if it is h-local and T_M has pseudo-Dedekind factorization for each maximal ideal M. By (2), R is h-local if and only if each S_α is h-local. Also, each maximal ideal N of S_α has the form $N = MS_\alpha$ for some maximal ideal of R and $(S_\alpha)_N = R_M$, and for each maximal ideal Q, $\text{supp}_{\mathscr{S}}(Q) = \{\beta\}$ for some $\beta \in \mathscr{A}$ and $R_Q = (S_\beta)_{QS_\beta}$ with $QS_\beta \in \text{Max}(S_\beta)$. Thus R_M has pseudo-Dedekind factorization for each maximal ideal M if and only if, for each $\alpha \in \mathscr{A}$, $(S_\alpha)_N$ has pseudo-Dedekind factorization for each maximal ideal N of S_α. Thus R has pseudo-Dedekind factorization if and only if each S_α has pseudo-Dedekind factorization.

For strong pseudo-Dedekind factorization, we simply use that a domain T has strong pseudo-Dedekind factorization if and only if it is an h-local Prüfer domain such that PT_P is a principal ideal of T_P for each nonzero prime ideal P of T. As with maximal ideals, each nonzero prime ideal of R survives in a unique S_α where it generates a prime ideal, and each nonzero prime ideal of each S_α is extended from a prime ideal of R. Moreover, $PR_P = P(S_\alpha)_{PS_\alpha}$ whenever $PS_\alpha \neq S_\alpha$. It follows that R has strong pseudo-Dedekind factorization if and only if each S_α has strong pseudo-Dedekind factorization. □

Theorem 6.3.9. *Let $\mathscr{S} := \{S_\alpha\}_{\alpha\in\mathscr{A}}$ be a Jaffard family with $R := \bigcap S_\alpha$. Also, for each $\alpha \in \mathscr{A}$, let S'_α denote the integral closure of S_α. Then the set $\mathscr{S}' := \{S'_\alpha\}_{\alpha\in\mathscr{A}}$ is a Jaffard family and $\bigcap_{\alpha\in\mathscr{A}} S'_\alpha$ is the integral closure of R.*

Proof. Let R' denote the integral closure of R. Then for each maximal ideal M of R, R'_M is the integral closure of R_M. Moreover, $R' = \bigcap\{R'_M \mid M \in \text{Max}(R)\}$. For each α, $(S_\alpha)_{MS_\alpha} = R_M$ for each $M \in \mathscr{X}_\alpha := \{M \in \text{Max}(R) \mid MS_\alpha \neq S_\alpha\}$

and $\mathrm{Max}(S_\alpha) = \{MS_\alpha \mid M \in \mathscr{X}_\alpha\}$ (Corollary 6.3.2). It follows that $S'_\alpha = \bigcap\{R'_M \mid M \in \mathscr{X}_\alpha\}$ and therefore $R' = \bigcap_{\alpha\in\mathscr{A}} S'_\alpha$.

Let N be a maximal ideal of R'. Then $M := N \cap R$ is a maximal ideal of R and $R'_M \subseteq R'_N$. Since $\mathscr{S}$ is a Jaffard family, $\mathrm{supp}_{\mathscr{S}}(M) = \{\beta\}$ for some $\beta \in \mathscr{A}$. It follows that $S'_\beta \subseteq R'_M \subseteq R'_N$. Thus N survives in S'_β. That $\mathscr{S}'$ is a Jaffard family now follows from Theorem 6.3.7. □

6.4 Factorization Examples

First a simple corollary to Theorem 2.5.10.

Corollary 6.4.1. *Let R be an integral domain and let $\mathscr{M}:=\{M_\alpha\}_{\alpha\in\mathscr{A}}$ be a nonempty set of maximal ideals of R. If $R = \bigcap_{\alpha\in\mathscr{A}} R_{M_\alpha}$, then each sharp maximal ideal of R (if any) is in the set $\mathscr{M}$, and $\bigcap_{\alpha\in\mathscr{A}} M_\alpha$ is the Jacobson radical of R.*

Proof. There is nothing to prove if $\mathscr{M} = \mathrm{Max}(R)$, so assume there is a maximal ideal M that is not one of the M_α's. Clearly, if $R = \bigcap_{\alpha\in\mathscr{A}} R_{M_\alpha}$, then M is not sharp. Moreover, for $t \in \bigcap_{\alpha\in\mathscr{A}} M_\alpha$, if t is not in M, then there are elements $p \in R$ and $q \in M$ such that $pt + q = 1$. It follows that q is a unit in each R_{M_α}. But having $R = \bigcap_{\alpha\in\mathscr{A}} R_{M_\alpha}$ implies q is a unit of R, a contradiction. Thus $t \in M$, and $\bigcap_{\alpha\in\mathscr{A}} M_\alpha$ is the Jacobson radical of R. □

We recall some results of Gilmer and Heinzer [36] concerning irredundant intersections of valuation domains.

First, a domain R is said to have an *S-representation* if there is a set of valuation overrings $\{V_\alpha\}_{\alpha\in\mathscr{A}}$ such that $R = \bigcap_{\alpha\in\mathscr{A}} V_\alpha$ and the intersection is irredundant (for each $\beta \in \mathscr{A}$, $R \subsetneq \bigcap_{\alpha\in\mathscr{A}\setminus\{\beta\}} V_\alpha$). Of course, in the case R is a Prüfer domain, each V_α is a localization of R. Moreover, as the intersection is irredundant, in the Prüfer case, each V_α is the localization at a sharp maximal ideal of R [36, Lemma 1.6], and each nonzero finitely generated ideal of R is contained in at least one such maximal ideal [36, Theorem 1.10].

Recall that a nonconstant polynomial $f(\mathrm{X}) \in R[\mathrm{X}]$ is said to be *unit valued* if $f(r)$ is a unit of R for each $r \in R$. The domain R is said to be a *non-D-ring* when such a polynomial exists (see, for example, [57, Page 271] and [39, Proposition 1]). If $R = \bigcap_{\alpha\in\mathscr{A}} V_\alpha$ where each V_α is a valuation domain and there is a polynomial $f(\mathrm{X}) \in R[\mathrm{X}]$, of degree 2 or more, that is unit valued in each V_α, then R is a Prüfer domain such that each invertible ideal has a power that is principal (see [57, Corollary 2.6]). Moreover, if there are two such polynomials where the gcd of their degrees is 1, then R is a Bézout domain [57, Corollary 2.7]. We make use of the latter result in several of our examples.

Throughout this section we let $G := \sum_{j=1}^{\infty} G_j$ under reverse lexicographic order where each $G_j := \mathbb{Z}$. Before presenting the examples, we set some notation that will be in effect for the first two examples of this section. Later more complicated examples will be created.

Start with a field K and let $\{X_{n,i}, Y \mid n, i \in \mathbb{Z}^+\}$ be a set of algebraically independent indeterminates over K. Also let $\mathcal{X} := \{X_{n,i} \mid n, i \in \mathbb{Z}^+\}$ and for each n, let $\mathcal{X}_n := \{X_{n,i} \mid i \in \mathbb{Z}^+\}$ and $\mathcal{X}_n^c := \mathcal{X} \setminus \mathcal{X}_n$. In the first two examples, we further assume that there is a subfield F of K such that $[K : F] = 2$. Also, we let $D := K[Y, \mathcal{X}]$ and $D^\flat := K(Y)[\mathcal{X}]$. For convenience we let $K_n := K(\mathcal{X}_n^c)$, $K_n^\flat := K(Y, \mathcal{X}_n^c)$, $D_n := K_n[\mathcal{X}_n]$ and $D_n^\flat := K_n[Y, \mathcal{X}_n]$.

Example 6.4.2. For each n, let V_n be the valuation overring of $D_n^\flat$ corresponding to the valuation map $v_n : K(Y, \mathcal{X}) \setminus \{0\} \to G$ defined as follows:

(a) $v_n(X_{n,i}) = e_i$ where e_i is the element of G all of whose components are 0 except the ith one which is a 1,
(b) $v_{n,i}(b) = 0$ for all nonzero $b \in K_n^\flat$,
(c) extend to all of $K(Y, \mathcal{X})$ using "min" for elements of $D_n^\flat$ (and sum for products and difference for fractions).

For each n, let N_n denote the maximal ideal of V_n. From the order on G, $N_n = X_{n,1}V_n$. Next, let $T := \bigcap_{n=1}^{\infty} V_n$. Then the following hold.

(1) T is a h-local Bézout domain with strong pseudo-Dedekind factorization.
(2) For each n, $M_n := N_n \cap T$ is a principal maximal ideal of T generated by $X_{n,1}$. These are the only maximal ideals of T.
(3) For each n, $T/M_n \cong V/N_n \cong K_n = K(Y, \mathcal{X}_n^c)$.
(4) For each n, $[K(Y, \mathcal{X}_n^c) : F(Y, \mathcal{X}_n^c)] = 2$.
(5) For each n, the pullback of $F(Y, \mathcal{X}_n^c)$ over M_n is a domain R_n with integral closure T that has pseudo-Dedekind factorization.

Proof. In this example, T is the Kronecker function ring that corresponds to the valuation domains $W_n := V_n \cap K(\mathcal{X})$. Thus T is a Bézout domain. For $m \neq n$, $X_{n,i}$ is a unit of V_m for each i. Thus $X_{n,i}^{-1} \in \bigcap_{m \neq n} V_m$ for each i. By [36, Lemma 1.6 & Theorem 1.10], each M_n is a maximal ideal of T and each finitely generated nonzero ideal of T is contained in at least one M_n.

Next, we show that not only is $M_n = X_{n,1}T$ for each n, but M_n is the only maximal ideal that contains $X_{n,i}$. After this we show that $\mathrm{Max}(T) = \{M_n \mid n \in \mathbb{Z}^+\}$. For the first part, let $h \in T \setminus M_n$ be a nonunit of T. Then h is a unit of V_n. Since T is an overring of $D^\flat$, $h = g/f$ where $g, f \in D^\flat$ have gcd 1 in $D^\flat$. For each positive integer k, $X_{n,i}^k + h$ is a unit of V_n. For sufficiently large k, each monomial term of $X_{n,i}^k f$ has total degree larger than the total degree of each monomial term of g. For such a k, no term of $X_{n,i}^k f$ cancels with a term of g. Thus $v_m(X_{n,i}^k f + g) = \min\{v_m(X_{n,i}^k f), v_m(g)\} = \min\{v_m(f), v_m(g)\} = v_m(f)$ for all $m \neq n$ (since $v_m(X_{n,i}) = 0$). It follows that $X_{n,i}^k + h$ is a unit in each V_k, and thus it is a unit of T. Therefore $M_n = X_{n,1}T$ is a maximal ideal of T and for each i, it is the only maximal ideal that contains $X_{n,i}$.

Next, we consider an arbitrary nonzero nonunit h of T. As above, we may write h as a quotient $h = g/f$ where $g, f \in D^\flat$ have gcd 1. Then h is in the base field of all but finitely many $D_n^\flat$s. Hence it is contained in at most finitely many M_ns. By [36, Corollary 1.11], these are the only maximal ideals of T that contain h.

Thus $\mathrm{Max}(T) = \{M_n \mid n \in \mathbb{Z}^+\}$ and it follows that each nonzero prime ideal is contained in a unique maximal ideal and therefore T is h-local. In addition, T has strong pseudo-Dedekind factorization since each maximal ideal is principal and PT_P is principal for each nonzero nonmaximal prime P.

Next, let F be a subfield of K such that $[K : F] = 2$ and choose a maximal ideal M_n (any will do). The corresponding residue field is $T/M_n = K(\mathrm{Y}, \mathscr{X}_n^c)$ and the subfield $F(\mathrm{Y}, \mathscr{X}_n^c)$ is such that $[K(\mathrm{Y}, \mathscr{X}_n^c) : F(\mathrm{Y}, \mathscr{X}_n^c)] = 2$. Thus by Theorem 6.1.6, the pullback of $F(\mathrm{Y}, \mathscr{X}_n^c)$ over M_n is a domain R_n with pseudo-Dedekind factorization that is not integrally closed. □

Remark 6.4.3. With the notation of the previous example, for each nonzero ideal A of R_n, there is a finite (nonempty) set of indeterminates $\{\mathrm{X}_{m_1,i_1}, \mathrm{X}_{m_2,i_2}, \ldots, \mathrm{X}_{m_r,i_r}\}$ and a positive integer k such that A contains the product $(\prod_{j=1}^r \mathrm{X}_{m_j,i_j})^k$. To see this first note since R_n is h-local, the ideal A has only finitely many minimal primes. By the construction of the V_ms, each minimal prime of R_n is the radical of some $\mathrm{X}_{m,i}$. Thus there is a finite set $\{\mathrm{X}_{m_1,i_1}, \mathrm{X}_{m_2,i_2}, \ldots, \mathrm{X}_{m_r,i_r}\}$ such that $\sqrt{A} = \sqrt{(\prod_{j=1}^r \mathrm{X}_{m_j,i_j})}$. Therefore there is a positive integer k such that $(\prod_{j=1}^r \mathrm{X}_{m_j,i_j})^k \in A$.

The domains in the next example have weak factorization but not strong. The larger, denoted T, is formed from the intersection of certain valuation overrings of the domain $D = K[\mathrm{Y}, \mathscr{X}]$ that all contain Y as a nonunit. The smaller ones, denoted R_n are formed by pullbacks, each with integral closure T.

Example 6.4.4. For each n, let V_n be the valuation overring of D_n corresponding to the valuation map $\mathrm{v}_n : K(\mathrm{Y}, \mathscr{X})\backslash\{0\} \to G$ defined as follows:

(a) $\mathrm{v}_n(\mathrm{X}_{n,i}) = e_i$ where e_i is the element of G all of whose components are 0 except the ith one which is a 1,
(b) $\mathrm{v}_{n,i}(b) = 0$ for all nonzero $b \in K_n$,
(c) there is a positive integer j_n such that $\mathrm{v}_n(\mathrm{Y}) := e_{j_n}$,
(d) extend to all of $K(\mathrm{Y}, \mathscr{X})$ using "min" for elements of D_n (and sum for products and difference for fractions).

For each n, let N_n denote the maximal ideal of V_n. From the order on G, $N_n = \mathrm{X}_{n,1}V_n$. Also by (a) and (c), $\mathrm{Y}/\mathrm{X}_{n,j_n}$ is a unit of V_n. Next, let $T := \bigcap_{n=1}^{\infty} V_n$. Then the following hold

(1) T is a Bézout domain.
(2) T is not h-local because Y is contained in infinitely many maximal ideals (in fact, it is in every maximal ideal).
(3) For each n, $M_n := N_n \cap T$ is a principal maximal ideal of T generated by $\mathrm{X}_{n,1}$. Moreover, $M := \mathrm{Y}K(\mathscr{X})[\mathrm{Y}]_{(\mathrm{Y})} \cap T$ is the only other maximal ideal of T and it has height one and is unsteady.
(4) T has weak factorization, but not strong factorization since it is not h-local.
(5) For each n, $T/M_n \cong V/N_n \cong K(\mathscr{X}_n^c, \mathrm{Y}/\mathrm{X}_{n,j_n})$.
(6) For each n, $[K(\mathscr{X}_n^c, \mathrm{Y}/\mathrm{X}_{n,j_n}) : F(\mathscr{X}_n^c, \mathrm{Y}/\mathrm{X}_{n,j_n})] = 2$.

(7) For each n, the pullback of $F(\mathscr{X}_n^c, \mathrm{Y}/\mathrm{X}_{n,j_n})$ over M_n is a domain R_n with integral closure T that has weak factorization.

Proof. Consider the polynomials $g_2(\mathrm{Z}) := \mathrm{Z}^2 + \mathrm{X}_{1,1} + \mathrm{X}_{2,1}$ and $g_3(\mathrm{Z}) := \mathrm{Z}^3 + \mathrm{X}_{1,1} + \mathrm{X}_{2,1}$ in $T[\mathrm{Z}]$, where Z is an indeterminate over the ring T. We will show that both are unit valued in each V_n. First, $g_2(0) = \mathrm{X}_{1,1} + \mathrm{X}_{2,1} = g_3(0)$ is a unit of V_n for each n. Next let $h/f \in V_n\backslash\{0\}$ where $h, f \in D$ and consider $g_2(h/f)$ and $g_3(h/f)$. Since $\mathrm{X}_{1,1} + \mathrm{X}_{2,1}$ is a unit of V_n, if h/f is a nonunit, then both $g_2(h/f)$ and $g_3(h/f)$ are units of V_n. So, we only need to consider what happens when h/f is a unit of V_n. The reasoning is similar for both $g_2(h/f)$ and $g_3(h/f)$. First rewrite each expression as a single fraction: $g_2(h/f) = (h^2+\mathrm{X}_{1,1}f^2+\mathrm{X}_{2,1}f^2)/f^2$ and $g_3(h/f) = (h^3 + \mathrm{X}_{1,1}f^3 + \mathrm{X}_{2,1}f^3)/f^3$. We have $\mathrm{v}_n(h) = \mathrm{v}_n(f)$. From the definition of v_n, the value $\mathrm{v}_n(h) = \mathrm{v}_n(f)$ is determined by the minimum value of the monomial terms of each polynomial. We may split each polynomial into sums $h = h_n + h_{n'}$ and $f = f_n + f_{n'}$ where h_n is the sum of the monomial terms of h with minimum value in V_n and f_n is the sum of the monomial terms of f with minimum value in V_n. Then $h^2 = h_n^2 + 2h_nh_{n'} + h_{n'}^2$, $h^3 = h_n^3 + 3h_n^2h_{n'} + 3h_nh_{n'}^2 + h_{n'}^3$, $f^2 = f_n^2 + 2f_nf_{n'} + f_{n'}^2$ and $f^3 = f_n^3 + 3f_n^2f_{n'} + 3f_nf_{n'}^2 + f_{n'}^3$. For both powers, the monomial terms with minimum value are those of h_n^2, h_n^3, f_n^2 and f_n^3.

Essentially, we have three cases to consider, $n = 1$, $n = 2$ and $n > 2$. No matter the value of n, at least one of $\mathrm{X}_{1,1}$ and $\mathrm{X}_{2,1}$ is a unit of V_n.

We start with the case $n > 2$. As both $\mathrm{X}_{1,1}$ and $\mathrm{X}_{2,1}$ are units of V_n, $\mathrm{v}_n(f_n^2) = \mathrm{v}_n(f_n^2\mathrm{X}_{1,1}) = \mathrm{v}_n(f_n^2\mathrm{X}_{2,1}) = \mathrm{v}_n(f_n^2(\mathrm{X}_{1,1} + \mathrm{X}_{2,1}))$. Moreover, each monomial term of $f_n^2\mathrm{X}_{1,1}$, $f_n^2\mathrm{X}_{2,1}$ and $f_n^2(\mathrm{X}_{1,1} + \mathrm{X}_{2,1})$ has this same value. It is clear that $h_n^2 + f_n^2(\mathrm{X}_{1,1} + \mathrm{X}_{2,1})$ cannot be the zero polynomial. Hence $\mathrm{v}_n(g_2(h/f)) = \mathrm{v}_n(h^2 + f^2(\mathrm{X}_{1,1} + \mathrm{X}_{2,1})) - \mathrm{v}_n(f^2) = \mathrm{v}_n(h_n^2 + f_n^2(\mathrm{X}_{1,1} + \mathrm{X}_{2,1})) - \mathrm{v}_n(f^2) = 0$ and thus $g_2(\mathrm{Z})$ is unit valued in V_n when $n > 2$. A similar proof shows that $g_3(\mathrm{Z})$ is also unit valued in V_n.

Next, consider the case $n = 1$. In V_1, $\mathrm{X}_{1,1}$ is a nonunit while $\mathrm{X}_{2,1}$ is a unit. Thus $\mathrm{v}_1(f_1^2\mathrm{X}_{1,1}) > \mathrm{v}_1(f_1^2) = \mathrm{v}_1(f_1^2\mathrm{X}_{2,1}) = \mathrm{v}_1(f_1^2(\mathrm{X}_{1,1} + \mathrm{X}_{2,1}))$. Moreover, each monomial term of $f_1^2\mathrm{X}_{2,1}$ has this same value, while all of the monomial terms of $f_1^2\mathrm{X}_{1,1}$ have larger value. It is clear that $h_1^2 + f_1^2\mathrm{X}_{1,1}$ cannot be the zero polynomial. Hence $\mathrm{v}_1(g_2(h/f)) = \mathrm{v}_1(h^2 + f^2(\mathrm{X}_{1,1} + \mathrm{X}_{2,1})) - \mathrm{v}_1(f^2) = \mathrm{v}_1(h_1^2 + f_1^2\mathrm{X}_{1,1}) - \mathrm{v}_1(f^2) = 0$ and thus $g_2(\mathrm{Z})$ is unit valued in V_1. Similar arguments show that $g_3(\mathrm{Z})$ is unit valued in V_1 and both $g_2(\mathrm{Z})$ and $g_3(\mathrm{Z})$ are unit valued in V_2. Therefore T is a Bézout domain by [57, Corollaries 2.6 & 2.7].

For each pair $n \neq m$, the sum $\mathrm{X}_{n,1} + \mathrm{X}_{m,1}$ is a unit in each V_k. Hence each such sum is a unit of T. Also, for each n, $M_n := N_n \cap T$ is a prime ideal of T that contains Y. Thus Y is contained in infinitely many maximal ideals of T and therefore T is not h-local. That Y is contained in the Jacobson radical will follow after we show that each M_n is a principal maximal ideal of T.

To see that each M_n is a principal maximal ideal of T, suppose $h \in T\backslash M_n$ is a nonunit of T. Then h is a unit of V_n. Since T is an overring of D, $h = g/f$ where $g, f \in D$ have gcd 1 in D. For each positive integer k, $\mathrm{X}_{n,1}^k + h$ is a unit of V_n. For sufficiently large k, each monomial term of $\mathrm{X}_{n,1}^kf$ has total degree larger than the total degree of each monomial term of g. For such a k, no term of $X_{n,1}^k f$ cancels with a term of g. Thus $v_m(X_{n,1}^k f + g) = \min\{v_m(X_{n,1}^k f), v_m(g)\} = \min\{v_m(f), v_m(g)\} = v_m(f)$ for all $m \neq n$ (since $v_m(X_{n,1}) = 0$). It follows that $X_{n,1}^k + h$ is a unit in each V_k, and thus a unit of T. Therefore M_n is a maximal ideal of T. This also shows that M_n is the only maximal ideal of T that contains $X_{n,1}$. Since we know T is a Bézout domain, it must be that $T_{M_n} = V_n$ with $M_n T_{M_n} = X_{n,1} V_n$. Hence we also have $M_n = X_{n,1} T$.

Similar analysis shows that for each $i \geq 1$, $X_{n,i}$ is comaximal with each element of $T \backslash M_n$. Hence M_n is the only maximal that contains $X_{n,i}$. It follows that M_n is the only maximal ideal of T that contains $\sqrt{X_{n,i} T}$. Thus $\sqrt{X_{n,i} T} = \sqrt{X_{n,i} V_n} \cap T$.

Let $\mathscr{A}$ be the set of finite products of distinct X_{n,j_n}s and consider the set $\mathscr{B} := \{Y/a \mid a \in \mathscr{A}\}$. Each finite subset of this set generates a proper ideal of all but finitely many V_ms. Hence this set generates a proper ideal B of T. On the other hand, no M_n contains the corresponding element $Y/X_{n,j_n}$ as this element is a unit of V_n. Thus no M_n contains B.

Let h be a nonzero element of $K[\mathscr{X}]$ that is not a unit of T. We will show that h is comaximal with B. If the constant term of h is nonzero, then h is a unit of each V_n and thus it is a unit of T. Hence the constant term of h is 0. For a given n, if some monomial term of h does not include a positive power of some $X_{n,i}$, then h is a unit of V_n. Thus h is a nonunit if and only if there is an integer n such that each monomial term of h includes a positive power of some $X_{n,i}$. As h is polynomial, there are at most finitely many such n, say $n_1, n_2, \ldots, n_r$. Let $g := \prod X_{n_i, j_{n_i}}$ and consider the sum $h + Y/g$.

For each n_i, Y/g is a unit of V_{n_i} while h is a nonunit. On the other hand, for $m \in \mathbb{Z}^+ \backslash \{n_1, n_2, \ldots, n_r\}$, h is a unit of V_m while Y/g is a nonunit. It follows that $h + Y/g$ is a unit in each V_n and thus is a unit of T.

Let P be a maximal ideal of T that contains B. Since each nonzero element of $K[\mathscr{X}]$ is comaximal with B, T_P contains $K(\mathscr{X})[Y]$. It must be that $T_P = K(\mathscr{X})[Y]_{(Y)}$ since $Y \in B$. Thus $P = M$ is unique and it is a height one maximal ideal of T.

Finally, suppose Q is a maximal ideal that is not one of the M_ns. Since Y is in the Jacobson radical of T, $Y \in Q$. For each n, M_n is the only maximal ideal that contains X_{n,j_n}. Thus having $X_{n,j_n} Y/X_{n,j_n} = Y$ in Q implies $Y/X_{n,j_n}$ is in Q. Similarly, we have $Y/a \in Q$ for each $a \in \mathscr{A}$. Therefore Q contains B and we have $Q = M$. It follows that M is the only other maximal ideal of T.

That M is not sharp follows from Theorem 2.5.10. On the other hand, from the previous observations we know that $MT_M = YK(\mathscr{X})[Y]_{(Y)}$ is principal. Thus M is an unsteady maximal ideal of T. All other maximal ideals (the M_ns) are principal. Also from the construction, each nonzero nonmaximal prime ideal is contained in unique maximal ideal and is sharp (in fact the radical of a principal ideal).

That T has weak factorization follows from Theorem 4.2.12.

Next, let F be a subfield of K such that $[K : F] = 2$ and choose a maximal ideal M_n (any will do). Since $Y/X_{n,j_n}$ is a unit of V_n, $T/M_n = K(\mathscr{X}_n^c, Y/X_{n,j_n})$. The subfield $F(\mathscr{X}_n^c, Y/X_{n,j_n})$ is such that $[K(\mathscr{X}_n^c, Y/X_{n,j_n}) : F(\mathscr{X}_n^c, Y/X_{n,j_n})] = 2$.

By Theorem 6.1.6, the pullback of $F(\mathscr{X}_n^c, \mathrm{Y}/\mathrm{X}_{n,j_n})$ over M_n is a domain R_n with weak factorization that is not integrally closed. □

Remark 6.4.5. The domains T and R_n from Example 6.4.4 have several other properties that we will find useful later.

(1) Since M_n is a common maximal ideal of R_n and T, $P_n := M \cap R_n$ is a maximal ideal of R_n and $(R_n)_{P_n} = T_M$. Moreover, for all $m \neq n$, $M_m \cap R_n = \mathrm{X}_{m,1} R_n$ is a principal maximal ideal of R_n, this follows easily from the fact that $(R_n)_{M_m \cap R_n} = T_{M_m}$ for all $m \neq n$. Also M_n is the radical of the principal ideal $\mathrm{X}_{n,1} R_n$.

(2) From the structure of the V_ms, each nonzero nonmaximal prime ideal Q of T is the radical of a principal ideal of the form $\mathrm{X}_{m,i} T$. Moreover, $Q \cap R_n$ is the radical of $\mathrm{X}_{m,i} R_n$. Since M_n is a common maximal ideal of R_n and T, each prime ideal of R_n is the contraction of a prime ideal of T. Thus except for P_n, each nonzero prime ideal of R_n is the radical of a principal ideal of the form $\mathrm{X}_{m,i} R_n$.

(3) If I is an ideal of R_n that is not contained in P_n, then IT is not contained in M. Moreover, since T is a Prüfer domain with weak factorization where M is the only unsteady maximal ideal, the ideal IT is contained in only finitely many maximal ideals of T (Corollary 4.2.6). It follows that IT has only finitely many minimal primes (Theorem 4.2.4). Thus I has only finitely many minimal primes in R_n.

(4) For each nonzero ideal A of R_n, there is a positive integer h and finitely many indeterminates $\mathrm{X}_{m_1,i_1}, \mathrm{X}_{m_2,i_2}, \ldots, \mathrm{X}_{m_s,i_s}$ such that the product $(\mathrm{Y} \prod_{j=1}^{s} \mathrm{X}_{m_j,i_j})^h$ is in A. To find such an element, first note that any finite set of $\mathrm{X}_{m,i}$s will do if A contains a positive power of Y. Thus we consider the case where A does not contain a positive power of Y. Since $(R_n)_{P_n} = K(\mathscr{X})[\mathrm{Y}]_{(\mathrm{Y})}$ is a discrete rank one valuation domain with maximal ideal generated by Y, $A(R_n)_{P_n} = \mathrm{Y}^k K(\mathscr{X})[\mathrm{Y}]_{(\mathrm{Y})}$ for some nonnegative integer k. It follows that the ideal $(A :_{R_n} \mathrm{Y}^k)$ is a proper ideal of R_n that is not contained in P_n. Thus $(A :_{R_n} \mathrm{Y}^k)$ has only finitely many minimal prime ideals, each of which is the radical of some $\mathrm{X}_{m,i}$. Let $\{\mathrm{X}_{m_1,i_1}, \mathrm{X}_{m_2,i_2}, \ldots, \mathrm{X}_{m_s,i_s}\}$ be the set of these indeterminates. Then $(A :_{R_n} \mathrm{Y}^k)$ has the same radical as the principal ideal $(\prod_{j=1}^{s} \mathrm{X}_{m_j,i_j}) R_n$. It follows that there is a positive integer $h \geq k$ such that $\mathrm{Y}^k (\prod_{j=1}^{s} \mathrm{X}_{m_j,i_j})^h \in A$. Hence $(\mathrm{Y} \prod_{j=1}^{s} \mathrm{X}_{m_j,i_j})^h$ is in A.

In addition to $G = \sum_{j=1}^{\infty} G_j$ as defined above, we let $H := \mathbb{R} \oplus G$ and $J := \mathbb{Q} \oplus G$, also under reverse lexicographic order.

In the next example, we let K be a field and let $\mathscr{Z} := \{\mathrm{Z}_{n,i} \mid n, i \in \mathbb{Z}^+\}$ be a set of algebraically independent indeterminates over K such that $\mathscr{Z} \cup \{\mathrm{Y}\}$ is also algebraically independent. We also let $D := K(\mathrm{Y})[\mathscr{R}]$ and $E := K(\mathrm{Y})[\mathscr{Q}]$ where $\mathscr{R} := \{\mathrm{Z}_{n,1}^{\alpha} \mid n \in \mathbb{Z}^+, \alpha \in \mathbb{R}^+\} \cup \{\mathrm{Z}_{n,j} \mid n, j \in \mathbb{Z}^+, 2 \leq j\}$ and $\mathscr{Q} := \{\mathrm{Z}_{n,1}^{\alpha} \mid n \in \mathbb{Z}^+, \alpha \in \mathbb{Q}^+\} \cup \{\mathrm{Z}_{n,j} \mid n, j \in \mathbb{Z}^+, 2 \leq j\}$. We will construct two Bézout domains using D and E as "base rings" much like we did in Example 6.4.2. The difference is that one of the domains will have pseudo-Dedekind factorization but not have strong pseudo-Dedekind factorization and the other will have strong factorization (equivalently, h-local) but not have pseudo-Dedekind factorization. Also, each domain has infinitely many idempotent maximal ideals and no other maximal ideals. As with D, we use different definitions for K_n and D_n. For each positive integer n, we let $K_n := K(\mathrm{Y}, \mathscr{R}_n^c)$, $D_n := K_n[\mathscr{R}_n]$, $L_n := K(\mathrm{Y}, \mathscr{Q}_n^c)$ and $E_n := L_n[\mathscr{Q}_n]$ where $\mathscr{R}_n := \{\mathrm{Z}_{n,1}^{\alpha} \mid \alpha \in \mathbb{R}^+\} \cup \{\mathrm{Z}_{n,j} \mid j \in \mathbb{Z}^+, 2 \leq j\}$, $\mathscr{R}_n^c := \mathscr{R} \backslash \mathscr{R}_n$, $\mathscr{Q}_n := \{\mathrm{Z}_{n,1}^{\alpha} \mid \alpha \in \mathbb{Q}^+\} \cup \{\mathrm{Z}_{n,j} \mid j \in \mathbb{Z}^+, 2 \leq j\}$ and $\mathscr{Q}_n^c := \mathscr{Q} \backslash \mathscr{Q}_n$.

Example 6.4.6. Let $D = K(\mathrm{Y})[\mathscr{R}]$ and $E = K(\mathrm{Y})[\mathscr{Q}]$ be as defined in the paragraph above (as well as K_n, L_n, D_n, E_n, etc.). For each positive integer n, define a pair of valuation domains V_n and W_n of $K(\mathrm{Y}, \mathscr{R})$ and $K(\mathrm{Y}, \mathscr{Q})$, respectively, corresponding to the valuations $\mathrm{v}_n : K(\mathrm{Y}, \mathscr{R}) \backslash \{0\} \to H$ and $\mathrm{w}_n : K(\mathrm{Y}, \mathscr{Q}) \backslash \{0\} \to J$ defined as follows.

(a) $\mathrm{v}_n(\mathrm{Z}_{n,1}^{\alpha}) = (\alpha, 0, 0, \ldots)$ and $\mathrm{w}_n(\mathrm{Z}_{n,1}^{\beta}) = (\beta, 0, 0, \ldots)$ for all $\alpha \in \mathbb{R}$ and $\beta \in \mathbb{Q}$.
(b) $\mathrm{v}_n(\mathrm{Z}_{n,j}) = (0, e_j) = \mathrm{w}_n(\mathrm{Z}_{n,j})$ for all $j \geq 2$ where $e_j \in G$ is the element of G whose jth coordinate is 1 and all others are 0 (as in the previous examples).
(c) $\mathrm{v}_n(b) = 0 = \mathrm{w}_n(c)$ for all nonzero $b \in K_n = K(\mathrm{Y}, \mathscr{R}_n^c)$ and $c \in L_n = K(\mathrm{Y}, \mathscr{Q}_n^c)$.
(d) Extend to valuations on $K(\mathrm{Y}, \mathscr{R})$ and $K(\mathrm{Y}, \mathscr{Q})$ using "min" for elements of $D_n = K_n[\mathscr{R}_n]$ and $E_n = L_n[\mathscr{Q}_n]$, respectively.

The domains $T := \bigcap_{n=1}^{\infty} V_n$ and $S := \bigcap_{n=1}^{\infty} W_n$ satisfy the following.

(1) $\mathrm{Max}(T) = \{N_n \cap T \mid n \in \mathbb{Z}^+\}$ and $\mathrm{Max}(S) = \{M_n \cap S \mid n \in \mathbb{Z}^+\}$ where N_n is the maximal ideal of V_n and M_n is the maximal ideal of W_n.
(2) Both T and S are Bézout domains with infinitely many maximal ideals (in fact, each is a Kronecker function ring). In addition, each maximal ideal is both branched and idempotent. Thus neither domain has strong pseudo-Dedekind factorization.
(3) Both T and S are h-local, so both have strong factorization.
(4) Since both T and S are h-local Bézout domains such that each maximal ideal is both branched and idempotent, if Q is a maximal ideal of either, then the intersection of the Q-primaries ideals is a prime ideal Q_0 that is properly contained in Q and contains all primes that are properly contained in Q (see Page 104).
(5) For each maximal ideal Q of T, the value group corresponding to $T/Q_0 = T_Q/Q_0T_Q$ is $\mathbb{R}$. Also PT_P is principal for each nonzero nonmaximal prime ideal P of T. Thus T has pseudo-Dedekind factorization.
(6) In contrast, for each maximal ideal Q of S, the value group corresponding to S/Q_0 ($= S_Q/Q_0S_Q$) is $\mathbb{Q}$. Thus S does not have pseudo-Dedekind factorization. However, PS_P is principal for each nonzero nonmaximal prime ideal P of S.

Proof. It is clear that V_n is the trivial extension in $K(\mathrm{Y}, \mathscr{R})$ of its contraction to $K(\mathscr{R})$ and W_n is the trivial extension in $K(\mathrm{Y}, \mathscr{Q})$ of its contraction to $K(\mathscr{Q})$. Thus both T and S are Kronecker function rings. It follows that each is a Bézout domain.

For each n, V_n and W_n are the only valuation domains in the corresponding family for which the $\mathrm{Z}_{n,i}$s are nonunits. It follows that each intersection is irredundant. Moreover, as in Example 6.4.2, each nonzero nonunit of T is in the base field for all but finitely many D_ns and the same holds for the nonzero nonunits of S with regard to the E_ns. Hence by [36, Corollary 1.11], each maximal ideal of T is the contraction of the maximal ideal of V_n for some unique n, and each maximal ideal of S is the contraction of the maximal ideal of W_n for some unique n. Moreover, we have that each nonzero nonmaximal prime ideal of T is contained in a unique maximal ideal, as is each nonzero nonmaximal prime ideal of S. Thus both T and S are h-local. If P is a nonzero nonmaximal prime ideal of either T or S, it is locally principal.

Let Q be a maximal ideal of T. Then $Q = N_n \cap T$ for some n, where N_n is the maximal ideal of V_n. Thus $T_Q = V_n$. Since the value group associated with V_n is $\mathbb{R} \oplus G$ under reverse lexicographic order, the value group associated with $V/N_{n,0}$ is $\mathbb{R}$. It follows that T has pseudo-Dedekind factorization.

In contrast, if Q is a maximal ideal of S, then the value group corresponding to W/M_n is $\mathbb{Q}$ (where $Q = M_n \cap S$ for some n). Hence S does not have pseudo-Dedekind factorization. □

The domains in the next three examples are formed by intersections of Jaffard families. Several of the domains have both infinitely many invertible maximal ideals and infinitely many idempotent maximal ideals. In each example, there are domains that are not integrally closed. Each of these has infinitely many divisorial maximal ideals that are neither invertible nor idempotent. The domains featured in Example 6.4.7 (the first of the three) have weak factorization but not strong factorization. In addition, each has infinitely many unsteady maximal ideals. In the second example (Example 6.4.8), the domains have pseudo-Dedekind factorization. Three of the domains featured in Example 6.4.9 have strong factorization but not pseudo-Dedekind factorization. In all of these examples, only the unsteady maximal ideals in the domains of the first have finite height. The constructions can be easily altered to produce maximal ideals of arbitrary finite heights.

As above start with a field K, but instead of one subfield F such that $[K : F] = 2$, we require that there is a subfield F of K and a countable set of elements $\mathscr{B} := \{\beta_n \mid n \in \mathbb{Z}^+\}$ in $K \backslash F$ such that $K = F(\mathscr{B})$ and for each n, $\beta_n^2 \in F$ and $\beta_n \notin F_n := F(\mathscr{B} \backslash \{\beta_n\})$. Next we let $\{\mathrm{X}_{n,i,j}, \mathrm{Z}_{n,j}, \mathrm{Y}_n \mid n, i, j \in \mathbb{Z}^+\}$ be a set of algebraically independent indeterminates over K.

In all three examples, we let $\mathscr{X} := \{\mathrm{X}_{n,i,j} \mid n,\ i,\ j \in \mathbb{Z}^+\}$ and $\mathscr{Y} := \{\mathrm{Y}_n \mid n \in \mathbb{Z}^+\}$. In the first two we make use of the set $\mathscr{R} = \{\mathrm{Z}_{n,1}^{\alpha} \mid \alpha \in \mathbb{R}^+\} \cup \{\mathrm{Z}_{n,j} \mid n, j \in \mathbb{Z}^+ \text{ with } 2 \leq j\}$ as defined earlier. In the third we use the set $\mathscr{Q} = \{\mathrm{Z}_{n,1}^{\alpha} \mid \alpha \in \mathbb{Q}^+\} \cup \{\mathrm{Z}_{n,j} \mid n, j \in \mathbb{Z}^+ \text{ with } 2 \leq j\}$. Note that the conclusions in Example 6.4.7 are equally valid if we use $\mathscr{Q}$ instead of $\mathscr{R}$.

For each n, we let $\mathscr{X}_n := \{\mathrm{X}_{n,i,j} \mid i, j \in \mathbb{Z}^+\}$, $\mathscr{X}_n^c := \mathscr{X} \backslash \mathscr{X}_n$ and $\mathscr{Y}_n^c := \mathscr{Y} \backslash \{\mathrm{Y}_n\}$. Also, for pairs n, i, we let $\mathscr{X}_{n,i} := \{\mathrm{X}_{n,i,j} \mid j \in \mathbb{Z}^+\}$ and $\mathscr{X}_{n,i}^c := \mathscr{X} \backslash \mathscr{X}_{n,i}$. In the first two examples, we make use of the sets $\mathscr{R}_n = \{\mathrm{Z}_{n,j} \mid 2 \leq j\} \cup \{\mathrm{Z}_{n,1}^{\alpha} \mid \alpha \in \mathbb{R}^+\}$ and $\mathscr{R}_n^c = \mathscr{Z} \backslash \mathscr{Z}_n$ as defined earlier, and in the third we use $\mathscr{Q}_n := \{\mathrm{Z}_{n,j} \mid 2 \leq j\} \cup \{\mathrm{Z}_{n,1}^{\alpha} \mid \alpha \in \mathbb{Q}^+\}$, $\mathscr{Q}_n^c := \mathscr{Z} \backslash \mathscr{Z}_n$

In Example 6.4.7, we let $D := K(\mathcal{R})[\mathcal{X}, \mathcal{Y}]$ and $E := K(\mathcal{X}, \mathcal{Y})[\mathcal{R}]$. For each n, let $K_n := K(\mathcal{R}_n^c, \mathcal{X}, \mathcal{Y})$ and $E_n := K_n[\mathcal{R}_n]$. In addition, for pairs n, i, we let $K_{n,i} := K(\mathcal{R}, \mathcal{X}_{n,i}^c, \mathcal{Y}_n^c)$, $D_{n,i} := K_{n,i}[\mathcal{X}_{n,i}, \mathrm{Y}_n]$ and $L_{n,i} := F_n(\mathcal{R}, \mathcal{X}_{n,i}^c, \mathcal{Y}_n^c)$.

As above, we have $G = \sum G_n$, $H = \mathbb{R} \oplus G$ and $J = \mathbb{Q} \oplus G$, all under reverse lexicographic order with each $G_n := \mathbb{Z}$.

Example 6.4.7. For each pair n, i, let $V_{n,i}$ be the valuation overring of $D_{n,i}$ corresponding to the valuation $\mathrm{v}_{n,i} : K(\mathcal{R}, \mathcal{X}, \mathcal{Y})\backslash\{0\} \to G$ defined as follows.

(a) $\mathrm{v}_{n,i}(\mathrm{X}_{n,i,j}) = e_j$ where e_j is the element of G all of whose components are 0 except the jth one which is a 1,
(b) $\mathrm{v}_{n,i}(b) = 0$ for all nonzero $b \in K_{n,i}$,
(c) there is a positive integer $j_{n,i}$ such that $\mathrm{v}_{n,i}(\mathrm{Y}_n) = e_{j_{n,i}}$,
(d) extend to all of $K(\mathcal{R}, \mathcal{X}, \mathcal{Y})$ using "min" for elements of $D_{n,i}$.

Also, for each positive integer n, let W_n be the valuation overring of E_n corresponding to the valuation $\mathrm{w}_n : K(\mathcal{R}, \mathcal{X}, \mathcal{Y})\backslash\{0\} \to H$ defined as follows.

(a) $\mathrm{w}_n(\mathrm{Z}_{n,1}^\alpha) = (\alpha, 0, 0, \dots)$ for all $\alpha \in \mathbb{R}$.
(b) $\mathrm{w}_n(\mathrm{Z}_{n,j}) = (0, e_j)$ for all $j \geq 2$ where $e_j \in G$ is the element of G whose jth coordinate is 1 and all others are 0 (as in the previous examples).
(c) $\mathrm{w}_n(b) = 0$ for all nonzero $b \in K_n = K(\mathcal{R}_n^c, \mathcal{X}, \mathcal{Y})$.
(d) Extend to valuations on $K(\mathcal{R}, \mathcal{X}, \mathcal{Y})$ using "min" for elements of E_n.

For each n, let $T_n := \bigcap_{i=1}^\infty V_{n,i}$. Each principal maximal ideal of T_n has the form $M_{n,i} := \mathrm{X}_{n,i,1}T_n$ for some pair n, i and for each pair n, i, the residue field $T_n/M_{n,i}$ is (naturally isomorphic to) $K_{n,i}(\mathrm{Y}_n/\mathrm{X}_{n,i,j_{n,i}})$. Also, let R_n be the pullback of $L_{n,1}(\mathrm{Y}_n/\mathrm{X}_{n,1,j_{n,1}})$ over $M_{n,1}$. Finally, let $T := \bigcap_{n=1}^\infty T_n$, $S := \bigcap_{n=1}^\infty W_n$, $R := \bigcap_{n=1}^\infty R_n$. Then the following hold.

(1) S is an h-local Bézout domain such that each maximal ideal is both branched and idempotent.
(2) For each n, T_n is a Bézout domain with weak factorization that has a unique unsteady maximal ideal P_n.
(3) For each n, R_n is a domain with integral closure T_n that has a unique divisorial maximal ideal that is neither invertible nor idempotent and a unique unsteady maximal ideal. All other maximal ideals of R_n are invertible, and R_n has weak factorization. Moreover, for each nonzero ideal A of R_n, $A \cap R \cap S$ contains an element that is a unit in all other R_ms and in all W_ks.
(4) $\{T_n \mid n \in \mathbb{Z}^+\}$, $\{T_n \mid n \in \mathbb{Z}^+\} \cup \{W_n \mid n \in \mathbb{Z}^+\}$, $\{R_n \mid n \in \mathbb{Z}^+\}$ and $\{R_n \mid n \in \mathbb{Z}^+\} \cup \{W_n \mid n \in \mathbb{Z}^+\}$ are Jaffard families.
(5) T is a Bézout domain with weak factorization that has infinitely many principal maximal ideals and infinitely many unsteady maximal ideals, but no idempotent maximal ideals.
(6) $S \cap T$ is a Bézout domain with weak factorization that has infinitely many principal maximal ideals, infinitely many unsteady maximal ideals and infinitely many idempotent maximal ideals.

(7) R is a domain with integral closure T that has weak factorization, infinitely many unsteady maximal ideals, infinitely many invertible maximal ideals and infinitely many maximal ideals that are divisorial but not invertible. It has no idempotent maximal ideals.
(8) $R \cap S$ is a domain with integral closure $S \cap T$ that has weak factorization, infinitely many unsteady maximal ideals, infinitely many invertible maximal ideals, infinitely many maximal ideals that are divisorial but not invertible and infinitely many idempotent maximal ideals.

Proof. For each positive integer n, we let $W_n^c := \bigcap_{m \neq n} W_m$, $R_n^c := \bigcap_{m \neq n} R_m$ and $T_n^c := \bigcap_{m \neq n} T_m$. Also, we let $\mathscr{R}_n^\flat := \bigcup_{j=1}^n \mathscr{R}_j$, $\mathscr{X}_n^\flat := \bigcup_{j=1}^n \mathscr{X}_j$ and $\mathscr{Y}_n^\flat := \{\mathrm{Y}_1, \mathrm{Y}_2, \ldots, \mathrm{Y}_n\}$.

By Example 6.4.6, S is an h-local Bézout domain with infinitely many idempotent maximal ideals, each of which is branched. Moreover, each maximal ideal of S is the contraction of the maximal ideal of some W_n. Thus the set $\{W_n\}$ is a Jaffard family and $\mathrm{Max}(S) = \{N_n \cap S \mid n \in \mathbb{Z}^+, N_n$ the maximal ideal of $W_n\}$.

By Example 6.4.4, each T_n is a Bézout domain with weak factorization (but not strong) and $\mathrm{Max}(T_n)$ is a countably infinite set. In addition, each T_n has a unique unsteady maximal ideal, P_n, and all other maximal ideals are principal. Since $[K_{n,1}(\mathrm{Y}_n/\mathrm{X}_{n,1,j_{n,1}}) : L_{n,1}(\mathrm{Y}_n/\mathrm{X}_{n,1,j_{n,1}})] = 2$, R_n has weak factorization (Theorem 6.1.6). From basic properties of pullbacks, T_n is the integral closure of R_n and $M_{n,1}$ is a divisorial maximal ideal of R_n that is not invertible. In addition, each maximal ideal of R_n is contracted from a maximal ideal of T_n, and if $h \in R$ is a unit of T_n, then it is also a unit of R_n. For $i > 1$, $M_{n,i,1} \cap R_n$ is an invertible maximal ideal of R_n, and $P_n \cap R$ is the unique unsteady maximal ideal of R_n. Also, from Remark 6.4.5(4), for each nonzero ideal A of R_n, there is a positive integer h and finitely many indeterminates, $\mathrm{X}_{n,i_1,j_1}, \mathrm{X}_{n,i_2,j_2}, \ldots, \mathrm{X}_{n,i_s,j_s}$ such that A contains the product $(\mathrm{Y}_n \prod_{k=1}^s \mathrm{X}_{n,i_k,j_k})^h$. The product $(\mathrm{Y}_n \prod_{k=1}^s \mathrm{X}_{n,i_k,j_k})^h \in A \cap R \cap S$ is a unit in all other R_ms and in all W_k.

For each n, β_n is not contained in R_n. Thus R does not contain $K[\mathscr{R}, \mathscr{X}, \mathscr{Y}]$. However, for each pair of positive integers $m \neq n$, $\beta_n \mathrm{Y}_n \in F_m[\mathscr{R}, \mathscr{X}, \mathscr{Y}] \subseteq R_m$. Also $\beta_n \mathrm{Y}_n \in M_{n,1}$, the common maximal ideal of R_n and T_n. Since each R_n contains $F[\mathscr{R}, \mathscr{X}, \mathscr{Y}]$, R contains $F[\mathscr{R}, \mathscr{X}, \mathscr{Y}, \{\beta_n \mathrm{Y}_n \mid n \in \mathbb{Z}^+\}]$. Also $\beta_n^2 \in F$ and $K = F[\beta]$. Therefore $K(\mathscr{R}, \mathscr{X}, \mathscr{Y})$ is the quotient field of R.

For each positive integer n, each nonzero element of $K(\mathscr{R}_n^\flat, \mathscr{X}_n^\flat, \mathscr{Y}_n^\flat) \cap R \cap S$ is a unit of all T_ms and W_ms for $m > n$. Thus each such element is a unit of R_m for all $m > n$. Since each nonzero element of $R \cap S$ is the quotient of a pair of nonzero elements of $K[\mathscr{R}, \mathscr{X}, \mathscr{Y}]$, $\{R_n \mid n \in \mathbb{Z}^+\}$ and $\{W_n \mid n \in \mathbb{Z}^+\}$ are families with finite character. Hence $|\mathrm{supp}(I)| < \infty$ for each nonzero ideal I of $R \cap S$. Also note that if $f \in R$ is a unit in each valuation domain $V_{n,i}$, then it is a unit of each R_n and thus a unit of R.

Next, we show that each proper ideal of $R \cap S$ survives in at least one R_n and/or at least one W_n.

Suppose $JW_j = W_j$ and $JR_j = R_j$ for all $j \leq k$. Then there is a finitely generated ideal $A \subseteq J$ such that $AW_j = W_j$ and $AR_j = R_j$ for all $1 \leq j \leq k$.

Thus $AV_{j,i} = V_{j,i}$ for all $1 \leq j \leq k$ and all i. Let $\{a_0, a_1, \ldots, a_r\}$ be a generating set for A. Then there is a positive integer $k' > k$ such that each a_i is in the field $K(\mathscr{R}^{\flat}_{k-1'}, \mathscr{X}^{\flat}_{k'-1}, \mathscr{Y}^{\flat}_{k'-1})$. Let $g := a_0 + a_1 \mathrm{Y}_{k'} + \cdots + a_r \mathrm{Y}^r_{k'}$. Then for each pair (m,i), $\mathrm{v}_{m,i}(g) = \min\{\mathrm{v}_{m,i}(a_0), \mathrm{v}_{m,i}(a_1) + \mathrm{v}_{m,i}(\mathrm{Y}_{k'}), \ldots, \mathrm{v}_{m,i}(a_r) + r\mathrm{v}_{m,i}(\mathrm{Y}_{k'})\}$. For $m > k$, $\mathrm{v}_{m,i}(a_0) = 0$. Thus $\mathrm{v}_{m,i}(g) = 0$ in this case. For $1 \leq k \leq m$, $\mathrm{v}_{m,i}(\mathrm{Y}_{k'}) = 0$ and $\mathrm{v}_{m,i}(a_j) = 0$ for some j, so again $\mathrm{v}_{m,i}(g) = 0$. Therefore g is in R. Since $\mathrm{w}_m(\mathrm{Y}_{k'}) = 0$ for all m, $\mathrm{w}_n(g) = \{\mathrm{w}_n(a_0), \mathrm{w}_n(a_1), \ldots, \mathrm{w}_n(a_r)\}$. As with the pairs (m,i), if $n > k$, then $\mathrm{w}_n(a_0) = 0$ and we have $\mathrm{w}_n(g) = 0$. On the other hand, if $1 \leq n \leq k$, then having $AW_n = W_n$ implies $\mathrm{w}_n(a_j) = 0$ for some j, so again $\mathrm{w}(g) = 0$ and we have that g is a unit of S. Hence g is a unit of $R \cap S$ and therefore $J = R \cap S$.

Therefore for each nonzero ideal I of $R \cap S$, $1 \leq |\mathrm{supp}(I)| \leq n$ for some positive integer n. Suppose I survives in W_m and let $J_m := IW_m \cap R \cap S$. From the definition of W_m, IW_m contains a positive power of some $\mathrm{Z}_{m,j}$ and thus so does J_m. Such an element is a unit in all other W_is and all R_ks. Hence $|\mathrm{supp}(J_m)| = 1$.

Next, suppose I survives in R_k for some k and let $B_k := IR_k \cap R \cap S$. Then B_k contains an element that is a unit in all other R_js and in all W_ms. Hence $|\mathrm{supp}(B_k)| = 1$. Thus $\{R_n \mid n \in \mathbb{Z}^+\} \cup \{W_n \mid n \in \mathbb{Z}^+\}$ is a Jaffard family by Theorem 6.3.5. By Theorem 6.3.7, $\{R_n \mid n \in \mathbb{Z}^+\}$ is also a Jaffard family. Since T_n is the integral closure of R_n for each n, $\{T_n \mid n \in \mathbb{Z}^+\} \cup \{W_n \mid n \in \mathbb{Z}^+\}$ and $\{R_n \mid n \in \mathbb{Z}^+\}$ are Jaffard families with T the integral closure of R and $T \cap S$ the integral closure of $R \cap S$ (Theorem 6.3.9).

By Theorem 6.3.8 and Example 6.4.4, both T and R have weak factorization, but neither has strong factorization. For each n, both T_n and R_n have a unique unsteady maximal ideal (of height one) and R_n has unique divisorial maximal ideal that is not invertible. All other maximal ideals of T_n and of R_n are invertible. Thus by Theorem 6.3.1, T has infinitely many unsteady maximal ideals and infinitely many invertible maximal ideals, but no idempotent maximal ideals. Also R has infinitely many unsteady maximal ideals, infinitely many invertible maximal ideals and infinitely many divisorial maximal ideals that are not invertible, but no idempotent maximal ideals.

Since S has pseudo-Dedekind factorization (Example 6.4.6), both $T \cap S$ and $R \cap S$ have weak factorization (Theorem 6.3.8 and Example 6.4.4). The domains S and T are independent as are S and R. Thus $\mathrm{Max}(S \cap T)$ is the disjoint union of $\{M \cap T \mid M \in \mathrm{Max}(S)\}$ and $\{N \cap S \mid N \in \mathrm{Max}(T)\}$ and $\mathrm{Max}(R \cap S)$ is the disjoint union of $\{M \cap R \mid M \in \mathrm{Max}(S)\}$ and $\{N \cap S \mid N \in \mathrm{Max}(R)\}$. Each maximal ideal in one of the intersections retains whatever special properties it had in the domain it came from. Thus $S \cap T$ has infinitely many branched idempotent maximal ideals, infinitely many invertible maximal ideals and infinitely many (height one) unsteady maximal ideals. Each maximal ideal of $R \cap S$ is contracted from a maximal ideal of $S \cap T$. Those that are contracted from unsteady maximal ideals retain that property, and those that are contracted from idempotent maximal ideals are idempotent. With regard to those that are contracted from invertible maximal ideals, infinitely many are invertible and infinitely many are divisorial but

not invertible. Specifically, $X_{n,1,1}T \cap R \cap S$ is a noninvertible maximal ideal of R that is divisorial, while $X_{n,i,1}T \cap R \cap S = X_{n,i,1}(R \cap S)$ is an invertible maximal ideal of $R \cap S$ for all $i > 1$.

By Theorem 6.3.8 both S and T are Prüfer domains. Using the same reasoning as that employed in the proof of Example 6.4.4, the polynomials $h_2(Z) = Z^2 + Y_1 + Y_2$ and $h_3(Z) = Z^3 + Y_1 + Y_2$ are unit valued in all W_ns and all $V_{n,i}$s. Hence both T and $S \cap T$ are Bézout domains [57, Corollary 2.7]. □

For Example 6.4.8 we make a slight change in the notation, we let $D := K(\mathscr{R}, \mathscr{Y})[\mathscr{X}]$ and $E := K(\mathscr{X}, \mathscr{Y})[\mathscr{R}]$. For each n, let $K_n := K(\mathscr{R}_n^c, \mathscr{X}, \mathscr{Y})$ and $E_n := K_n[\mathscr{R}_n]$. In addition, for pairs n, i, we let $K_{n,i} := K(\mathscr{R}, \mathscr{X}_{n,i}^c, \mathscr{Y})$, $D_{n,i} := K_{n,i}[\mathscr{X}_{n,i}]$ and $L_{n,i} := F_n(\mathscr{R}, \mathscr{X}_{n,i}^c, \mathscr{Y})$. As above, $K = F[\beta_1, \beta_2, \ldots]$ such that for each n, $\beta_n \in \bigcap_{m \neq n} F_m \setminus F_n$, $K = F_n[\beta_n]$ and $\beta_n^2 \in F$. The valuation domains W_n will be defined as in Example 6.4.7, but for the $V_{n,i}$s we map all Y_ks to 0, and thus avoid having unsteady maximal ideals.

Example 6.4.8. For each pair n, i, let $V_{n,i}$ be the valuation overring of $D_{n,i}$ corresponding to the valuation $v_{n,i} : K(\mathscr{R}, \mathscr{X}, \mathscr{Y}) \setminus \{0\} \to G$ defined as follows.

(a) $v_{n,i}(X_{n,i,j}) = e_j$ where e_j is the element of G all of whose components are 0 except the jth one which is a 1,
(b) $v_{n,i}(b) = 0$ for all nonzero $b \in K_{n,i}$,
(c) extend to all of $K(\mathscr{R}, \mathscr{X}, \mathscr{Y})$ using "min" for elements of $D_{n,i}$.

Also, for each positive integer n, let W_n be the valuation overring of E_n corresponding to the valuation $w_n : K(\mathscr{R}, \mathscr{X}, \mathscr{Y}) \setminus \{0\} \to H$ defined as follows.

(a) $w_n(Z_{n,1}^{\alpha}) = (\alpha, 0, 0, \ldots)$ for all $\alpha \in \mathbb{R}$.
(b) $w_n(Z_{n,j}) = (0, e_j)$ for all $j \geq 2$ where $e_j \in G$ is the element of G whose jth coordinate is 1 and all others are 0 (as in the previous examples).
(c) $w_n(b) = 0$ for all nonzero $b \in K_n = K(\mathscr{R}_n^c, \mathscr{X}, \mathscr{Y})$.
(d) Extend to valuations on $K(\mathscr{R}, \mathscr{X}, \mathscr{Y})$ using "min" for elements of E_n.

For each n, let $T_n := \bigcap_{i=1}^{\infty} V_{n,i}$. Each principal maximal ideal of T_n has the form $M_{n,i} := X_{n,i,1}T_n$ for some pair n, i and for each pair n, i, the residue field $T_n/M_{n,i}$ is (naturally isomorphic to) $K_{n,i}$. Also let R_n be the pullback of $L_{n,1}$ over $M_{n,1}$. Finally, let $T := \bigcap_{n=1}^{\infty} T_n$, $S := \bigcap_{n=1}^{\infty} W_n$, $R := \bigcap_{n=1}^{\infty} R_n$. Then the following hold.

(1) S is an h-local Bézout domain with pseudo-Dedekind factorization such that each maximal ideal is both branched and idempotent.
(2) For each n, T_n is an h-local Bézout domain such that each maximal ideal is principal. In addition, T_n has strong pseudo-Dedekind factorization.
(3) For each n, R_n is a domain with integral closure T_n that has a unique divisorial maximal ideal that is neither invertible nor idempotent. All other maximal ideals of R_n are invertible, and R_n has pseudo-Dedekind factorization. Moreover, for each nonzero ideal A of R_n, $A \cap R \cap S$ contains an element that is a unit in all other R_ms and in all W_ks.

(4) $\{T_n \mid n \in \mathbb{Z}^+\}$, $\{T_n \mid n \in \mathbb{Z}^+\} \cup \{W_n \mid n \in \mathbb{Z}^+\}$, $\{R_n \mid n \in \mathbb{Z}^+\}$ and $\{R_n \mid n \in \mathbb{Z}^+\} \cup \{W_n \mid n \in \mathbb{Z}^+\}$ are Jaffard families.
(5) T is an h-local Bézout domain with strong pseudo-Dedekind factorization. In addition, T has infinitely many maximal ideals each of which is principal.
(6) $S \cap T$ is an h-local Bézout domain with strong pseudo-Dedekind factorization weak factorization that has infinitely many branched idempotent maximal ideals and infinitely many invertible maximal ideals.
(7) R is an h-local domain with integral closure T that has pseudo-Dedekind factorization. It has infinitely many divisorial maximal ideals that are not invertible and infinitely many invertible maximal ideals, but no idempotent maximal ideals.
(8) $R \cap S$ is an h-local domain with integral closure $T \cap S$ that has pseudo-Dedekind factorization. It has infinitely many divisorial maximal ideals that are neither invertible nor idempotent, infinitely many invertible maximal ideals and infinitely many branched idempotent maximal ideals.

Proof. Each $V_{n,i}$ is the trivial extension to $K(\mathscr{R}, \mathscr{X})(\mathscr{Y})$ of its contraction to $K(\mathscr{R}, \mathscr{X})$. The same is true for each W_n. Thus T, S and $S \cap T$ are all Kronecker function rings, and so each is a Bézout domain. By Example 6.4.6, S is h-local with pseudo-Dedekind factorization. As in Example 6.4.7, each nonzero nonunit of $S \cap T$ is a unit in all but finitely many T_ns and finitely many W_ns. In addition, each T_n is h-local with strong pseudo-Dedekind factorization (Example 6.4.2). It follows that each nonzero nonunit of $S \cap T$ is a unit in all but finitely many $V_{n,i}$s and W_ms. Since $(R_n : T_n)$ is common maximal ideal of R_n and T_n, if $g \in R_n$ is a unit of T_n, then it is a unit of R_n.

Let A be a nonzero ideal of R_n. By Remark 6.4.3, there is a finite set $\{\mathrm{X}_{n,i_j,k_j} \mid 1 \leq j \leq r\}$ and a positive integer h such that $(\prod_{j=1}^{r} \mathrm{X}_{n,i_j,k_j})^h \in A$. The product $(\prod_{j=1}^{r} \mathrm{X}_{n,i_j,k_j})^h$ is a unit in S and in all other R_ms.

From the construction of the R_ns, each contains $F[\mathscr{R}, \mathscr{X}, \mathscr{Y}]$ as does S. Thus $R \cap S \supseteq F[\mathscr{R}, \mathscr{X}, \mathscr{Y}]$. In addition, S contains $\beta_n \mathrm{X}_{n,1,1}$ for each n. For $m \neq n$, $\beta_n \in F_m \subseteq R_m$ and $\beta_n \mathrm{X}_{n,1,1} \in M_{n,1} \subseteq R_n$. Thus $\beta_n \mathrm{X}_{n,1,1} \in R$ for each n. Therefore the ring $R \cap S$ contains $F[\mathscr{R}, \mathscr{X}, \mathscr{Y}, \{\beta_n \mathrm{X}_{n,1,1} \mid n \in \mathbb{Z}^+\}]$. Since $K = F[\beta_1, \beta_2, \ldots]$ and F contains β_n^2 for each n, $K(\mathscr{R}, \mathscr{X}, \mathscr{Y})$ is the quotient field of $R \cap S$.

For each nonzero element $f \in R \cap S$, there is a positive integer k and a finite set of pairs $\{(m_1, i_1), (m_2, i_2), \ldots, (m_h, i_h)\}$ such that $f \in K(\mathscr{R}_k^\flat, \bigcup_{j=1}^{h} \mathscr{X}_{m_j,i_j}, \mathscr{Y}_k^\flat)$. Then f is a unit of W_m for each integer $m > k$ and f is a unit of $V_{n,i}$ for each pair $(n, i) \notin \{(m_1, i_1), (m_2, i_2), \ldots, (m_h, i_h)\}$. It follows that f is a unit in all but finitely many R_ns.

Next we show that each proper ideal of $R \cap S$ survives in at least one $V_{n,i}$ and/or at least one W_n.

Let J be a nonzero ideal of $R \cap S$ and let a_0 be a nonzero element of J. Then there is a positive integer k and a finite set of pairs $\{(m_1, i_1), (m_2, i_2), \ldots, (m_h, i_h)\}$ such that $a_0 \in K(\mathscr{R}_k^\flat, \bigcup_{j=1}^{h} \mathscr{X}_{m_j,i_j}, \mathscr{Y}_k^\flat)$. For each integer $m > k$, a_0 is a unit of W_m. Also, a_0 is a unit of $V_{n,i}$ for each pair $(n, i) \notin \{(m_1, i_1), (m_2, i_2), \ldots, (m_h, i_h)\}$.

Thus $JW_m = W_m$ for each $m > k$ and $JV_{n,i} = V_{n,i}$ for each $(n,i) \notin \{(m_1,i_1),(m_2,i_2),\ldots,(m_h,i_h)\}$.

Suppose $JW_j = W_j$ for all $j \leq k$ and $JR_{m_g,i_g} = R_{m_g,i_g}$ for all pairs (m_g,i_g) belonging to $\{(m_1,i_1),(m_2,i_2),\ldots,(m_h,i_h)\}$. Then there is a finitely generated ideal $A \subseteq J$ such that $AR_{m_g,i_g} = R_{m_g,i_g}$ for all pairs $(m_g,i_g) \in \{(m_1,i_1),(m_2,i_2),\ldots,(m_h,i_h)\}$ and $AW_j = W_j$ for all $1 \leq j \leq k$. Let $\{a_0,a_1,\ldots,a_r\}$ be a generating set for the ideal A. Then there is a pair of positive integers $h' \geq h$ and $k' > k$ and a finite set of pairs $\{(m_1,i_1),(m_2,i_2),\ldots,(m_{h'},i_{h'})\}$ ($\supseteq \{(m_1,i_1),(m_2,i_2),\ldots, (m_h,i_h)\}$) such that each a_i is in the field $K(\mathscr{R}^{\flat}_{k'-1}, \bigcup_{j=1}^{h'} \mathscr{X}_{m_j,i_j}, \mathscr{Y}^{\flat}_{k'-1})$. As in the proof of Example 6.4.7, we let $g := a_0 + a_1\mathrm{Y}_{k'} + \cdots + a_r\mathrm{Y}^r_{k'}$. Unlike what happens in Example 6.4.7, $\mathrm{Y}_{k'}$ is a unit in all $V_{m,i}$s and all W_ms. Thus for each pair of positive integers (m,i), $\mathrm{v}_{m,i}(g) = \min\{\mathrm{v}_{m,i}(a_0),\mathrm{v}_{m,i}(a_1),\ldots,\mathrm{v}_{m,i}(a_r)\}$. For $(m,i) \notin \{(m_1,i_1),(m_2,i_2),\ldots, (m_h,i_h)\}$, $\mathrm{v}_{m,i}(a_0) = 0$; and for $(m,i) \in (m_1,i_1),(m_2,i_2),\ldots,(m_h,i_h)\}$, $\mathrm{v}_{m,i}(a_j) = 0$ for some j. Thus g is a unit of each $V_{m,i}$. Similarly, $\mathrm{w}_m(g) = \min\{\mathrm{w}_m(a_0),\mathrm{w}_m(a_1),\ldots,\mathrm{w}_m(a_r)\}$ for each positive integer m, with $\mathrm{w}_m(a_0) = 0$ for $m > k$, and $\mathrm{w}_m(a_j) = 0$ for some j when $1 \leq m \leq k$. Thus g is a unit of each W_m. As $g \in R \cap S$, it is a unit of $R \cap S$. Therefore $J = R \cap S$.

For a proper ideal I of $R \cap S$, $1 \leq |\mathrm{supp}(I)| < \infty$. Moreover, if $IR_n \neq R_n$, then $IR_n \cap R \cap S$ contains an element that is a unit in all other R_ms and in all W_ms. Similarly, if $IW_m \neq W_m$ for some m, then $IW_m \cap R \cap S$ contains a positive power of some $\mathrm{Z}_{m,i}$, an element which is a unit in all other W_ks and all R_ns. Therefore the family $\{R_n \mid n \in \mathbb{Z}^+\} \cup \{W_n \mid n \in \mathbb{Z}^+\}$ is a Jaffard family by Theorem 6.3.5. By Theorems 6.3.7 and Theorems 6.3.9, $\{R_n \mid n \in \mathbb{Z}^+\}$, $\{T_n \mid n \in \mathbb{Z}^+\} \cup \{W_n \mid n \in \mathbb{Z}^+\}$ and $\{T_n \mid n \in \mathbb{Z}^+\}$ are Jaffard families with T the integral closure of R and $T \cap S$ the integral closure of $R \cap S$.

Each maximal ideal of R is contracted from a maximal ideal of some unique R_n. Each R_n has a unique divisorial maximal ideal that is not invertible (the conductor of T_n into R_n) and all others are invertible (in fact, principal). By Theorem 6.3.1, each invertible maximal ideal of R_n contracts to an invertible maximal ideal of R, and the divisorial maximal ideal that is not invertible contracts to a divisorial maximal ideal of R that is not invertible. Each of these maximal ideals contracts to a maximal ideal of $R \cap S$ of the same type. In addition, each maximal ideal of S contracts to a branched idempotent maximal ideal of $R \cap S$. There are no other maximal ideals of $R \cap S$.

By Theorem 6.3.8, both R and $R \cap S$ have pseudo-Dedekind factorization. □

For Example 6.4.9 we make yet another change in the notation. Essentially, we simply use $\mathscr{Q}$, $\mathscr{Q}_n$ and $\mathscr{Q}^c_n$ in place of $\mathscr{R}$, $\mathscr{R}_n$ and $\mathscr{R}^c_n$. Specifically, we let $D := K(\mathscr{Q},\mathscr{Y})[\mathscr{X}]$ and $E := K(\mathscr{X},\mathscr{Y})[\mathscr{Q}]$. For each n, let $K_n := K(\mathscr{Q}^c_n,\mathscr{X},\mathscr{Y})$ and $E_n := K_n[\mathscr{Q}_n]$. In addition, for pairs n,i, we let $K_{n,i} := K(\mathscr{Q},\mathscr{X}^c_{n,i},\mathscr{Y})$, $D_{n,i} := K_{n,i}[\mathscr{X}_{n,i}]$ and $L_{n,i} := F_n(\mathscr{Q},\mathscr{X}^c_{n,i},\mathscr{Y})$. The W_ns will be defined in the same way as in Example 6.4.8 except now the corresponding value group will be $J = \mathbb{Q} \oplus G$ (instead of $\mathbb{R} \oplus G$).

Example 6.4.9. For each pair n, i, let $V_{n,i}$ be the valuation overring of $D_{n,i}$ corresponding to the valuation $\mathrm{v}_{n,i} : K(\mathcal{Q}, \mathcal{X}, \mathcal{Y})\backslash\{0\} \to G$ defined as follows.

(a) $\mathrm{v}_{n,i}(\mathrm{X}_{n,i,j}) = e_j$ where e_j is the element of G all of whose components are 0 except the jth one which is a 1,
(b) $\mathrm{v}_{n,i}(b) = 0$ for all nonzero $b \in K_{n,i}$,
(c) extend to all of $K(\mathcal{Q}, \mathcal{X}, \mathcal{Y})$ using "min" for elements of $D_{n,i}$.

Also for each positive integer n, let W_n be the valuation overring of E_n corresponding the valuation $\mathrm{w}_n : K(\mathcal{Q}, \mathcal{X}, \mathcal{Y})\backslash\{0\} \to J$ defined as follows.

(a) $\mathrm{w}_n(\mathrm{Z}_{n,1}^{\alpha}) = (\alpha, 0, 0, \ldots)$ for all $\alpha \in \mathbb{R}$.
(b) $\mathrm{w}_n(\mathrm{Z}_{n,j}) = (0, e_j)$ for all $j \geq 2$ where $e_j \in G$ is the element of G whose jth coordinate is 1 and all others are 0 (as in the previous examples).
(c) $\mathrm{w}_n(b) = 0$ for all nonzero $b \in K_n = K(\mathcal{Q}_n^c, \mathcal{X}, \mathcal{Y})$.
(d) Extend to valuations on $K(\mathcal{Q}, \mathcal{X}, \mathcal{Y})$ using "min" for elements of E_n.

For each n, let $T_n := \bigcap_{i=1}^{\infty} V_{n,i}$. Each principal maximal ideal of T_n has the form $M_{n,i} := \mathrm{X}_{n,i,1}T_n$ for some pair n, i and for each pair n, i, the residue field $T_n/M_{n,i}$ is (naturally isomorphic to) $K_{n,i}$. Also, let R_n be the pullback of $L_{n,1}$ over $M_{n,1}$. Finally, let $T := \bigcap_{n=1}^{\infty} T_n$, $S := \bigcap_{n=1}^{\infty} W_n$, $R := \bigcap_{n=1}^{\infty} R_n$. Then the following hold.

(1) S is an h-local Bézout domain such that each maximal ideal is both branched and idempotent. Also S has strong factorization but not pseudo-Dedekind factorization.
(2) For each n, T_n is an h-local Bézout domain such that each maximal ideal is principal.
(3) For each n, R_n is a domain with integral closure T_n that has a unique divisorial maximal ideal that is neither invertible nor idempotent. All other maximal ideals of R_n are invertible, and R_n has pseudo-Dedekind factorization. Moreover, for each nonzero ideal A of R_n, $A \cap R \cap S$ contains an element that is a unit in all other R_ms and in all W_ks.
(4) $\{T_n \mid n \in \mathbb{Z}^+\}$, $\{T_n \mid n \in \mathbb{Z}^+\} \cup \{W_n \mid n \in \mathbb{Z}^+\}$, $\{R_n \mid n \in \mathbb{Z}^+\}$ and $\{R_n \mid n \in \mathbb{Z}^+\} \cup \{W_n \mid n \in \mathbb{Z}^+\}$ are Jaffard families.
(5) T is an h-local Bézout domain with strong pseudo-Dedekind factorization and infinitely many maximal ideals such that each maximal ideal is invertible.
(6) $S \cap T$ is an h-local Bézout domain with strong factorization but not pseudo-Dedekind factorization. In addition, $S \cap T$ has infinitely many branched idempotent maximal ideals and infinitely many invertible maximal ideals.
(7) R is a domain with integral closure T that has pseudo-Dedekind factorization. It has infinitely many divisorial maximal ideals that are not invertible and infinitely many invertible maximal ideals, but no idempotent maximal ideals.
(8) $R \cap S$ is a domain with integral closure $T \cap S$ that has strong factorization but not pseudo-Dedekind factorization. It has infinitely many divisorial maximal ideals that are neither invertible nor idempotent, infinitely many invertible maximal ideals and infinitely many branched idempotent maximal ideals.

Proof. Essentially, one may repeat the proof of Example 6.4.8. The only differences are with regard to pseudo-Dedekind factorization. For R and T there are no changes, both still have pseudo-Dedekind factorization and the description of the maximal ideals stays the same. While the basic description of the maximal ideals of S and $R \cap S$ stays the same, here the domain S has strong factorization but not pseudo-Dedekind factorization (by Example 6.4.6). Hence $R \cap S$ has strong factorization but not pseudo-Dedekind factorization. □

Symbols and Definitions

Max(R, I)**:** The set of maximal ideals of the domain R that contain the ideal I. (Page 10)

Max$^{\#}(R)$**:** The set of sharp maximal ideals of R. (Page 45)

Max$^{\dagger}(R)$**:** The set of dull maximal ideals of R. (Page 46)

Min(R, I)**:** The set of minimal primes of the ideal I in the domain R. (Page 11)

supp$_{\mathscr{S}}(I)$**:** The set $\{\beta \in \mathscr{A} \mid IS_\beta \neq S_\beta\}$ where $\mathscr{S} = \{S_\alpha\}_{\alpha \in \mathscr{A}}$ is a family of domains with the same quotient field and I is an ideal of $\bigcap_{\alpha \in \mathscr{A}} S_\alpha$. The support of I with respect to $\mathscr{S}$. (Page 129)

supp$_{\mathscr{P}}(I)$**:** The set $\{\beta \in \mathscr{A} \mid IW_\beta \neq W_\beta\}$ where I is an ideal of a domain R, $\mathscr{P} = \{\mathscr{X}_\alpha\}_{\alpha \in \mathscr{A}}$ is a partition of Max(R) and $W_\alpha = \bigcap\{R_M \mid M \in \mathscr{X}_\alpha\}$ for each $\alpha \in \mathscr{A}$. The support of I with respect to $\mathscr{P}$. (Page 131)

$\Gamma_R(I)$**:** The ring $\bigcap\{R_M \mid M \in \mathrm{Max}(R, I)\}$. (Page 10)

$\Gamma(I)$**:** Same as $\Gamma_R(I)$, used when R is understood to be the relevant domain. (Page 10)

$\mathscr{H}(I)$**:** The (possibly empty) set of maximal ideals M of an integral domain R such that $IR_M \neq (IR_M)^v$. (Page 72)

$\Phi_R(I)$**:** The ring $\bigcap\{R_P \mid P \in \mathrm{Min}(R, I)\}$. (Page 11)

$\Phi(I)$**:** Same as $\Phi_R(I)$, used when R is understood to be the relevant domain. (Page 11)

$\Theta_R(I)$**:** The ring $\bigcap\{R_N \mid N \in \mathrm{Max}(R)\backslash\mathrm{Max}(R, I)\}$, equal to the quotient field when each maximal ideal contains I. (Page 10)

$\Theta(I)$**:** Same as $\Theta_R(I)$, used when R is understood to be the relevant domain. (Page 10)

M. Fontana et al., *Factoring Ideals in Integral Domains*, Lecture Notes of the Unione Matematica Italiana 14, DOI 10.1007/978-3-642-31712-5,

$\Omega_R(I)$: The ring $\bigcap\{R_P \mid P \in \text{Spec}(R)\ ,\ P \not\supseteq I\}$, equal to the quotient field when each maximal ideal contains I. (Page 11)

$\Omega(I)$: Same as $\Omega_R(I)$, used when R is understood to be the relevant domain. (Page 11)

$c(h)$: The ideal of R generated by the coefficients of the polynomial $h \in R[X]$, referred to as the content h. (Page 36)

$c(I)$: The ideal of R generated by the contents of the polynomials in an ideal I of the polynomial ring $R[X]$, referred to as the content of I. Also used for an ideal I of $R(X)$; in this case, $I = L_I R(X)$ for the ideal $L_I = I \bigcap R[X]$ and $\mathbf{c}(I) = \mathbf{c}(L_I)$. (Page 114)

#-domain: A domain R with the property that for each pair of nonempty subsets Δ' and Δ'' of Max(R), $\Delta' \neq \Delta''$ implies $\bigcap\{R_{M'} \mid M' \in \Delta'\} \neq \bigcap\{R_{M''} \mid M'' \in \Delta''\}$. (Page 9)

##-domain: A domain such that every overring is a #-domain. (Page 9)

Almost Dedekind domain: A domain R such that for each maximal ideal M R_M is a rank-one discrete valuation domain (or, equivalently, a (local) Dedekind domain). (Page 9)

Anneau du type de Dedekind: A domain for which each nonzero ideal can be factored as a finite product of ideals with each factor in a unique and distinct maximal ideal. The same as an h-local domain. (Page 7)

Antesharp prime: A nonzero prime ideal P of a domain R is antesharp if each maximal ideal of $(P : P)$ that contains P contracts to P in R. (Page 19)

aRTP-domain: A domain R such that for each nonzero noninvertible I, $II^{-1}R_M$ is a radical ideal whenever M is either a steady maximal ideal or an unsteady maximal ideal that is not minimal over II^{-1}. Alternately, one may say that R has the almost radical trace property. (Page 26)

Branched prime: A prime ideal P that has proper P-primary ideals. (Page 13)

Content of h: For a polynomial $h \in R[X]$, the content of h, $\mathbf{c}(h)$, is the ideal of R generated by the coefficients of h. (Page 36)

Content of I: For a nonzero ideal I of $R[X]$, the content of I, $\mathbf{c}(I)$, is the ideal of R generated by the coefficients of the polynomials in I. In the case I is an ideal of $R(X)$, then the content of I is $\mathbf{c}(I) = \mathbf{c}(L_I)$ where $L_I = I \bigcap R[X]$. (Page 114)

Dull degree: For a one-dimensional Prüfer domain R with quotient field K, recursively define a family of overrings of R as follows:

$$R_1 := R, \text{ and } R_n := \bigcap\{(R_{n-1})_N \mid N \in \text{Max}^\dagger(R_{n-1})\} \text{ for } n > 1.$$

Then R has dull degree n, if $R_{n-1} \subsetneq R_n = R_{n+1} \subsetneq K$ (with $R_0 = \{0\}$). (Page 46)

Dull maximal ideal: It is a nonsharp maximal ideal of a one-dimensional Prüfer domain. (Page 46)

Factoring family: For an almost Dedekind domain R, a family of finitely generated ideals $\{J_\alpha \mid J_\alpha R_{M_\alpha} = M_\alpha R_{M_\alpha}, M_\alpha \in \text{Max}(R)\}$ such that each finitely generated nonzero ideal can be factored as a finite product of powers of ideals from the family with negative exponents allowed. (Page 46)

Factoring set: For an almost Dedekind domain, it is a factoring family such that no member appears more than once. (Page 46)

Finite character: A domain for which each nonzero nonunit is contained in only finitely many maximal ideals. (Page 6)

Finite divisorial closure property: A domain for which each nonzero nondivisorial ideal I has the property that $I^v = I + J$ for some finitely generated ideal J. (Page 88)

Finite idempotent character: A domain for which each nonzero nonunit is contained in at most finitely many idempotent maximal ideals. (Page 77)

Finite unsteady character: A domain for which each nonzero nonunit is contained in at most finitely many unsteady maximal ideals. (Page 77)

Generalized Dedekind domain: A Prüfer domain such that each nonzero prime ideal is the radical of a finitely generated ideal and no nonzero prime is idempotent. (Page 47)

General ZPI-ring: A ring for which each proper ideal factors as a finite product of prime ideals. Alternately one may say R is a general "Zerlegung PrimIdeale" ring. For domains, same as Dedekind domain and ZPI-ring. (Page 6)

h-local domain: A domain such that each nonzero prime ideal is contained in a unique maximal ideal and each nonzero ideal (equivalently, each nonzero nonunit) is contained in only finitely many maximal ideals. (Page 6)

h-local maximal ideal: A maximal ideal M of a domain R such that $R_M \Theta_R(M)$ is the quotient field of R. (Page 120)

Independent pair of domains: A pair of domains S and T with the same quotient field K such that $ST = K$ and no nonzero prime ideal of $S \cap T$ survives in both S and T. (Page 126)

Jaffard family: A family of domains $\mathcal{S} = \{S_\alpha\}_{\alpha \in \mathcal{A}}$ with the same quotient field K such that $R = \bigcap_{\alpha \in \mathcal{A}} S_\alpha$ also has quotient field K and for each nonzero ideal I of R: $\text{supp}_{\mathcal{S}}(I) = \{\alpha_1, \alpha_2, \ldots, \alpha_n\}$ some positive integer n, $I = \prod_{i=1}^n (I \cap S_{\alpha_i})$, and $IS_{\alpha_i} \cap R$ and $IS_{\alpha_j} \cap R$ are comaximal for all $1 \le i < j \le n$. (Page 129)

Locally pseudo-valuation domain: A domain R such that R_M is a pseudo-valuation domain for each maximal ideal M of R. (Page 107)

Matlis partition: A partition $\mathscr{P} = \{X_\alpha\}_{\alpha\in\mathscr{A}}$ of Max(R) such that $|\text{supp}_{\mathscr{P}}(P)| = 1$ for each nonzero prime ideal P of R and $|\text{supp}_{\mathscr{P}}(rR)| < \infty$ for each nonzero nonunit $r \in R$. (Page 131)

Non-D-ring: A domain R with a polynomial $f(x) \in R[x]\backslash R$ such that $f(r)$ is a unit of R for all $r \in R$ ($f(x)$ is a unit valued polynomial). (Page 136)

Prestable ideal: An ideal I of a ring R (not necessarily a domain) such that, for each prime P, there is a positive integer n such that $I^{2n}R_P = dI^nR_P$ for some $d \in I^n$. (Page 100)

Principally complete valuation domain: A valuation domain V with the following property: whenever there are two families of nonzero elements $\{b_\alpha\}_{\alpha\in\mathscr{A}}$ and $\{c_\alpha\}_{\alpha\in\mathscr{A}}$ and a corresponding family of primes $\{P_\alpha\}_{\alpha\in\mathscr{A}}$ with $\mathscr{A}$ totally ordered such that for all $\alpha < \beta$ in $\mathscr{A}$:

(i) $b_\alpha \in P_\alpha$,
(ii) $b_\alpha V \subseteq b_\beta V \subseteq c_\beta V \subseteq c_\alpha V$,
(iii) $b_\alpha/c_\alpha \in V\backslash P_\alpha$, and
(iv) $P_\alpha \subseteq P_\beta$ with $\bigcup P_\alpha$ an unbranched prime,

then there is an element $c \in V$ such that $b_\alpha V \subseteq cV \subseteq c_\alpha V$ for all $\alpha \in \mathscr{A}$. (Page 104)

Property (α)**:** An integral domain R is said to have *property* (α) if every primary ideal is a power of its radical. (Page 41)

Pseudo-Dedekind factorization: A factorization of a nonzero ideal as the product of an invertible ideal and a finite product of pairwise comaximal prime ideals with at least one prime in the second factor. A domain has pseudo-Dedekind factorization if each nonzero noninvertible ideal has a pseudo-Dedekind factorization. (Page 95)

Pseudo-valuation domain: A local domain (R, M) such that $(M : M)$ is a valuation domain with maximal ideal M. Same as PVD. (Page 101)

PVD: Same as pseudo-valuation domain. (Page 101)

Radical factorization: A domain for which each nonzero ideal I factors as $I = I_1I_2\cdots I_n$ for some radical ideals $I_1, I_2, \cdots, I_n$. Same as a SP-domain. (Page 39)

Reflexive domain: A domain R such that $\text{Hom}_R(-, R)$ induces a duality on submodules of finite rank of free R-modules. (Page 8)

Relatively sharp prime: A nonzero prime ideal P of a Prüfer domain is relatively sharp in a nonempty set of incomparable primes $\mathscr{S}$ if P contains a (nonzero) finitely generated ideal that is contained in no other prime in the set $\mathscr{S}$. (Page 34)

Relatively sharp set: A nonempty set $\mathscr{S}$ consisting of incomparable nonzero primes of a Prüfer domain such that each prime in the set is relatively sharp in $\mathscr{S}$. (Page 34)

RTP-domain: A domain R with the property that II^{-1} is a radical ideal for each nonzero noninvertible I. Alternately, one may say that R has the radical trace property. (Page 20)

Sharp degree: For a one-dimensional Prüfer domain R with quotient field K, recursively define a family of overrings of R as follows:

$$R_1 := R, \text{ and } R_n := \bigcap\{(R_{n-1})_N \mid N \in \mathrm{Max}^\dagger(R_{n-1})\} \text{ for } n > 1,$$

where $R_n = K$, for $n \geq 2$, if $\mathrm{Max}^\dagger(R_{n-1}) = \emptyset$. Then R has sharp degree n, if $R_n \neq K$ but $R_{n+1} = K$. In addition, a fractional ideal J of R has sharp degree n if $JR_n \neq R_n$ but $JR_{n+1} = R_{n+1}$. (Page 46)

Sharp prime: A nonzero prime P of a domain R such that R_P does not contain $\Theta(P)$. (Page 10)

SP-domain: A domain with radical factorization. (Page 39)

Special factorization: A special factorization of a nonzero ideal I of a domain R is a factorization of the form $I = BP_1P_2 \cdots P_n$ for some finitely generated ideal B and (not necessarily distinct) prime ideals $P_1, P_2, \ldots, P_n$. A domain R has special factorization if each nonzero ideal has a special factorization. For R, this is equivalent to it having strong pseudo-Dedekind factorization; and to it being a ZPUI-ring. (Page 108)

S-representation: A set of valuation overrings $\{V_\alpha\}_{\alpha \in \mathscr{A}}$ of a domain R such that $R = \bigcap_{\alpha \in \mathscr{A}} V_\alpha$ is an irredundant intersection. (Page 136)

Stable ideal: An ideal I of a ring R such that for each prime ideal P of R, $I^2R_P = dIR_P$ for some $d \in R$. (Page 100)

Steady/Unsteady maximal ideal: A maximal ideal M of a domain R is unsteady if MR_M is principal but $M(R : M) \neq R$, otherwise M is steady. (Page 8)

Strongly discrete Prüfer domain: A Prüfer domain R such that PR_P is principal for each nonzero prime ideal P. (Page 47)

Strongly discrete valuation domain: A valuation domain V such that PV_P is principal for each nonzero prime ideal P. (Page 47)

Strong pseudo-Dedekind factorization: A domain R for which each nonzero ideal has a pseudo-Dedekind factorization. Equivalent to special factorization for R; and to R being a ZPUI-ring. (Page 96)

Strong factorization: A domain for which each nonzero nondivisorial ideal I factors as $I = I^vM_1M_2 \cdots M_n$ for some distinct maximal ideals $M_1, M_2, \ldots, M_n$. (Page 72)

SV-stable (stable in the sense of Sally-Vasconcelos): an ideal I which is invertible as ideal of $(I : I)$. (Page 37)

TP-domain: A domain R with the property that II^{-1} is a prime ideal for each nonzero noninvertible ideal I. Alternately, one may say that R has the trace property. (Page 20)

TPP-domain: A domain R with the property that QQ^{-1} is a prime ideal for each nonzero noninvertible ideal primary ideal Q. Alternately, one may say that R has the trace property for primary ideals. (Page 23)

Trace ideal: An ideal J of the form $J = II^{-1}$ for some nonzero ideal I. Also, the trace of a nonzero ideal B is the ideal BB^{-1}. (Page 20)

Unbranched prime: A prime ideal P that has no proper P-primary ideals. (Page 13)

Unit valued polynomial: A polynomial $f(x) \in R[x]\backslash R$ such that $f(r)$ is a unit of R for each $r \in R$. (Page 136)

Very strong factorization: A domain R for which each nonzero nondivisorial ideal I factors as $I = I^v M_1 M_2 \cdots M_n$ where the M_is are the distinct nondivisorial maximal ideals that contain I where IR_{M_i} is not divisorial. (Page 72)

Weak factorization: A domain for which each nonzero nondivisorial ideal I factors as $I^v M_1^{r_1} M_2^{r_2} \cdots M_n^{r_n}$ for some maximal ideals $M_1, M_2, \ldots, M_n$. (Page 71)

wTPP-domain: A domain R for which $QQ^{-1} = \sqrt{Q}$ for each nonzero primary ideal Q with nonmaximal radical. Alternately one may say that R has the weak trace property for primary ideals. (Page 26)

ZPI-ring: A ring for which each nonzero ideal factors as a unique finite product of prime ideals. Alternately, one may say the ring is a "Zerlegung Primideale" ring. For domains, same as Dedekind domain and general ZPI-ring. (Page 6)

ZPUI-ring: A ring R for which each nonzero ideal I factors as a product $I = BP_1 P_2 \cdots P_n$ for some invertible ideal B (possibly equal to R) and prime ideals $P_1, P_2, \ldots, P_n$ with $n \geq 1$. Alternately, one may say R is a "Zerlegung Prim- und Umkehrbaridealen" ring. For domains, the same as a domain with strong pseudo-Dedekind factorization; and a domain with special factorization. (Page 95)

References

[1] D.D. Anderson, Non-finitely generated ideals in valuation domains. Tamkang J. Math. **18**, 49–52 (1987)

[2] D.D. Anderson, J. Huckaba, I. Papick, A note on stable domains. Houst. J. Math. **13**, 13–17 (1987)

[3] D.D. Anderson, L. Mahaney, Commutative rings in which every ideal is a product of primary ideals. J. Algebra **106**, 528–535 (1987)

[4] D.D. Anderson, M. Zafrullah, Independent locally-finite intersections of localizations. Houst. J. Math. **25**, 433–452 (1999)

[5] D.F. Anderson, D.E. Dobbs, M. Fontana, On treed Nagata rings. J. Pure Appl. Algebra **61**, 107–122 (1989)

[6] E. Bastida, R. Gilmer, Overrings and divisorial ideals of rings of the form $D + M$. Mich. Math. J. **20**, 79–95 (1973)

[7] S. Bazzoni, Class semigroups of Prüfer domains. J. Algebra **184**, 613–631 (1996)

[8] S. Bazzoni, Clifford regular domains. J. Algebra **238**, 703–722 (2001)

[9] S. Bazzoni, L. Salce, Warfield domains. J. Algebra **185**, 836–868 (1996)

[10] N. Bourbaki, *Algèbre Commutative*, Chapitres I et II (Hermann, Paris, 1961)

[11] J. Brewer, W. Heinzer, On decomposing ideals into products of comaximal ideals. Comm. Algebra **30**, 5999–6010 (2002)

[12] H.S. Butts, Unique factorization of ideals into nonfactorable ideals. Proc. Am. Math. Soc. **15**, 21 (1964)

[13] H.S. Butts, R. Gilmer, Primary ideals and prime power ideals. Can. J. Math. **18**, 1183–1195 (1966)

[14] H.S. Butts, R.W. Yeagy, Finite bases for integral closures. J. Reine Angew. Math. **282**, 114–125 (1976)

[15] P.-J. Cahen, T. Lucas, The special trace property, in *Commutative Ring Theory* (Fès, Morocco, 1995), Lecture Notes in Pure and Applied Mathematics, vol. 185 (Marcel Dekker, New York, 1997), pp. 161–172

[16] D.E. Dobbs, M. Fontana, Locally pseudo-valuation domains. Ann. Mat. Pura Appl. **134**, 147–168 (1983)

[17] P. Eakin, A. Sathaye, Prestable ideals. J. Algebra **41**, 439–454 (1976)

[18] S. El Baghdadi, S. Gabelli, Ring-theoretic properties of PvMDs. Comm. Algebra **35**, 1607–1625 (2007)

[19] M. Fontana, E. Houston, T. Lucas, Factoring ideals in Prüfer domains. J. Pure Appl. Algebra **211**, 1–13 (2007)

[20] M. Fontana, E. Houston, T. Lucas, Toward a classification of prime ideals in Prüfer domains. Forum Math. **22**, 741–766 (2010)

M. Fontana et al., *Factoring Ideals in Integral Domains*, Lecture Notes of the Unione Matematica Italiana 14, DOI 10.1007/978-3-642-31712-5,

[21] M. Fontana, J. Huckaba, I. Papick, Divisorial prime ideals in Prüfer domains. Can. Math. Bull. **27**, 324–328 (1984)

[22] M. Fontana, J. Huckaba, I. Papick, Some properties of divisorial prime ideals in Prüfer domains. J. Pure Appl. Algebra **39**, 95–103 (1986)

[23] M. Fontana, J. Huckaba, I. Papick, Domains satisfying the trace property. J. Algebra **107**, 169–182 (1987)

[24] M. Fontana, J. Huckaba, I. Papick, *Prüfer Domains* (Marcel Dekker, New York, 1997)

[25] M. Fontana, J. Huckaba, I. Papick, M. Roitman, Prüfer domains and endomorphism rings of their ideals. J. Algebra **157**, 489–516 (1993)

[26] M. Fontana, N. Popescu, Sur une classe d'anneaux qui généralisent les anneaux de Dedekind. J. Algebra **173**, 44–66 (1995)

[27] L. Fuchs, L. Salce, *Modules Over Non-Noetherian Domains*, Math. Surveys & Monographs, vol. 84 (American Mathematical Society, Providence, 2001)

[28] S. Gabelli, A class of Prüfer domains with nice divisorial ideals, in *Commutative Ring Theory* (Fès, Morocco, 1995), Lecture Notes in Pure and Applied Mathematics, vol. 185 (Marcel Dekker, New York, 1997), pp. 313–318

[29] S. Gabelli, Generalized Dedekind domains, in *Multiplicative Ideal Theory in Commutative Algebra* (Springer, New York, 2006), pp. 189–206

[30] S. Gabelli, E. Houston, Coherent-like conditions in pullbacks. Mich. Math. J. **44**, 99–123 (1997)

[31] S. Gabelli, N. Popescu, Invertible and divisorial ideals of generalized Dedekind domains. J. Pure Appl. Algebra **135**, 237–251 (1999)

[32] R. Gilmer, Integral domains which are almost Dedekind. Proc. Am. Math. Soc. **15**, 813–818 (1964)

[33] R. Gilmer, Overrings of Prüfer domains. J. Algebra **4**, 331–340 (1966)

[34] R. Gilmer, *Multiplicative Ideal Theory*, Queen's Papers in Pure and Applied Mathematics, vol. 90 (Queen's University Press, Kingston, 1992)

[35] R. Gilmer, W. Heinzer, Overrings of Prüfer domains. II. J. Algebra **7**, 281–302 (1967)

[36] R. Gilmer, W. Heinzer, Irredundant intersections of valuation domains. Math. Z. **103**, 306–317 (1968)

[37] R. Gilmer, J. Hoffmann, A characterization of Prüfer domains in terms of polynomials. Pac. J. Math. **60**, 81–85 (1975)

[38] R. Gilmer, J. Huckaba, The transform formula for ideals. J. Algebra **21**, 191–215 (1972)

[39] H. Gunji, D.L. McQuillan, On rings with a certain divisibility property. Mich. Math. J. **22**, 289–299 (1975)

[40] F. Halter-Koch, Kronecker function rings and generalized integral closures. Comm. Algebra **31**, 45–59 (2003)

[41] F. Halter-Koch, Clifford semigroups of ideals in monoids and domains. Forum Math. **21**, 1001–1020 (2009)

[42] J.H. Hays, The S-transform and the ideal transform. J. Algebra **57**, 223–229 (1979)

[43] J. Hedstrom, E. Houston, Pseudo-valuation domains. Pac. J. Math. **75**, 137–147 (1978)

[44] W. Heinzer, Integral domains in which each non-zero ideal is divisorial. Mathematika **15**, 164–170 (1968)

[45] W. Heinzer, J. Ohm, Locally Noetherian commutative rings. Trans. Am. Math. Soc. **158**, 273–284 (1971)

[46] W. Heinzer, I. Papick, The radical trace property. J. Algebra **112**, 110–121 (1988)

[47] M. Henriksen, On the prime ideals of the ring of entire functions. Pac. J. Math. **3**, 711–720 (1953)

[48] E. Houston, S.-E. Kabbaj, T. Lucas, A. Mimouni, When is the dual of an ideal a ring? J. Algebra **225**, 429–450 (2000)

[49] W.C. Holland, J. Martinez, W.Wm. McGovern, M. Tesemma, Bazzoni's conjecture. J. Algebra **320**, 1764–1768 (2008)

[50] J. Huckaba, in *Commutative Rings with Zero Divisors*, Pure and Applied Mathematics, vol. 117 (Marcel Dekker, New York, 1988)

[51] J. Huckaba, I. Papick, When the dual of an ideal is a ring? Manuscripta Math. **37**, 67–85 (1982)
[52] P. Jaffard, Théorie arithmétique des anneau du type de Dedekind. Bull. Soc. Math. France **80**, 61–100 (1952)
[53] C. Jayram, Almost Q-rings. Arch. Math. (Brno) **40**, 249–257 (2004)
[54] I. Kaplansky, *Commutative Rings* (Allyn and Bacon, Boston, 1970)
[55] K. Kubo, Über die Noetherschen fünf Axiome in kommutativen Ringen. J. Sci. Hiroshima Univ. Ser. A **10**, 77–84 (1940)
[56] J. Lipman, Stable rings and Arf rings. Am. J. Math. **93**, 649–685 (1971)
[57] K.A. Loper, On Prüfer non-D-rings. J. Pure Appl. Algebra **96**, 271–278 (1994)
[58] A. Loper, T. Lucas, Factoring ideals in almost Dedekind domains. J. Reine Angew. Math. **565**, 61–78 (2003)
[59] T. Lucas, The radical trace property and primary ideals. J. Algebra **184**, 1093–1112 (1996)
[60] E. Matlis, *Cotorsion Modules*, vol. 49 (Memoirs of the American Mathematical Society, Providence RI, 1964)
[61] E. Matlis, Reflexive domains. J. Algebra **8**, 1–33 (1968)
[62] K. Matsusita, Über ein bewertungstheoretisches Axiomensystem für die Dedekind-Noethersche Idealtheorie. Jpn. J. Math. **19**, 97–110 (1944)
[63] W.Wm. McGovern, Prüfer domains with Clifford class semigroup. J. Comm. Algebra **3**, 551–559 (2011)
[64] S. Mori, Allgemeine Z.P.I.-Ringe. J. Sci. Hiroshima Univ. A **10**, 117–136 (1940)
[65] N. Nakano, Idealtheorie in einem speziellen unendlichen algebraischen Zahlkörper. J. Sci. Hiroshima Univ. A **16**, 425–439 (1953)
[66] E. Noether, Abstrakter Aufbau der Idealtheorie in algebraischen Zahl- und Funktionenkörpern. Math. Ann. **96**, 26–61 (1927)
[67] B. Olberding, Globalizing local properties of Prüfer domains. J. Algebra **205**, 480–504 (1998)
[68] B. Olberding, Factorization into prime and invertible ideals. J. Lond. Math. Soc. **62**, 336–344 (2000)
[69] B. Olberding, Factorization into radical ideals, in *Arithmetical Properties of Commutative Rings and Monoids*, Lecture Notes Pure and Applied Mathematics, vol. 241 (Chapman & Hall/CRC, Boca Raton, 2005), pp. 363–377
[70] B. Olberding, Factorization into prime and invertible ideals, II. J. Lond. Math. Soc. **80**, 155–170 (2009)
[71] B. Olberding, Characterizations and constructions of h-local domains, in *Contributions to Module Theory* (Walter de Gruyter, Berlin, 2008), pp. 385–406
[72] E. Popescu, N. Popescu, A characterization of generalized Dedekind domains. Bull. Math. Roumanie **35**, 139–141 (1991)
[73] N. Popescu, On a class of Prüfer domains. Rev. Roumaine Math. Pures Appl. **29**, 777–786 (1984)
[74] F. Richman, Generalized quotient rings. Proc. Am. Math. Soc. **16**, 794–799 (1965)
[75] N.H. Vaughan, R.W. Yeagy, Factoring ideals in semiprime ideals. Can. J. Math. **30**, 1313–1318 (1978)
[76] C.A. Wood, On general Z.P.I. rings. Pac. J. Math. **30**, 837–846 (1969)
[77] R.W. Yeagy, Semiprime factorizations in unions of Dedekind domains. J. Reine Angew. Math. **310**, 182–186 (1979)
[78] M. Zafrullah, t-Invertibility and Bazzoni-like statements. J. Pure Appl. Algebra **214**, 654–657 (2010)

Index

M. Fontana et al., *Factoring Ideals in Integral Domains*, Lecture Notes of the Unione Matematica Italiana 14, DOI 10.1007/978-3-642-31712-5,

Editor in Chief: Franco Brezzi

Editorial Policy

1. The UMI Lecture Notes aim to report new developments in all areas of mathematics and their applications - quickly, informally and at a high level. Mathematical texts analysing new developments in modelling and numerical simulation are also welcome.

2. Manuscripts should be submitted (preferably in duplicate) to
   Redazione Lecture Notes U.M.I.
   Dipartimento di Matematica
   Piazza Porta S. Donato 5
   I – 40126 Bologna
   and possibly to one of the editors of the Board informing, in this case, the Redazione about the submission. In general, manuscripts will be sent out to external referees for evaluation. If a decision cannot yet be reached on the basis of the first 2 reports, further referees may be contacted. The author will be informed of this. A final decision to publish can be made only on the basis of the complete manuscript, however a refereeing process leading to a preliminary decision can be based on a pre-final or incomplete manuscript. The strict minimum amount of material that will be considered should include a detailed outline describing the planned contents of each chapter, a bibliography and several sample chapters.

3. Manuscripts should in general be submitted in English. Final manuscripts should contain at least 100 pages of mathematical text and should always include

   – a table of contents;
   – an informative introduction, with adequate motivation and perhaps some historical remarks: it should be accessible to a reader not intimately familiar with the topic treated;
   – a subject index: as a rule this is genuinely helpful for the reader.

4. For evaluation purposes, manuscripts may be submitted in print or electronic form (print form is still preferred by most referees), in the latter case preferably as pdf- or zipped ps-files. Authors are asked, if their manuscript is accepted for publication, to use the LaTeX2e style files available from Springer's web-server at
   ftp://ftp.springer.de/pub/tex/latex/svmonot1/ for monographs
   and at
   ftp://ftp.springer.de/pub/tex/latex/svmultt1/ for multi-authored volumes

5. Authors receive a total of 50 free copies of their volume, but no royalties. They are entitled to a discount of 33.3% on the price of Springer books purchased for their personal use, if ordering directly from Springer.

6. Commitment to publish is made by letter of intent rather than by signing a formal contract. Springer-Verlag secures the copyright for each volume. Authors are free to reuse material contained in their LNM volumes in later publications: A brief written (or e-mail) request for formal permission is sufficient.